RECENT TRENDS IN MULTI-USER MIMO COMMUNICATIONS

Edited by **Maha Ben Zid**

Recent Trends in Multi-user MIMO Communications
http://dx.doi.org/10.5772/3306
Edited by Maha Ben Zid

Contributors

Giovanni Geraci, Jinhong Yuan, Rizwan Ghaffar, Elli Kartsakli, Nizar Zorba, Luis Alonso, Christos Verikoukis, Maha Ben Zid, Kosai Raoof, Hieu Nguyen, Rui Zhang, Hon Tat Hui, Qiaoyu Li, Jinho Choi, Lin Bai, Ying Li, Wei Peng, Fumiyuki Adachi

Published by InTech

Janeza Trdine 9, 51000 Rijeka, Croatia

Notice

Publishing Process Manager Sandra Bakic
Technical Editor InTech DTP team
Cover InTech Design team

Additional hard copies can be obtained from orders@intechopen.com

Recent Trends in Multi-user MIMO Communications, Edited by Maha Ben Zid
p. cm.
ISBN 978-953-51-1210-5

For EU product safety concerns:
IN TECH d.o.o., Prolaz Marije Krucifikse Kozulić 3, 51000 Rijeka, Croatia,
info@intechopen.com or visit our website at intechopen.com.

Meet the editor

Maha BEN ZID is an associate Professor at the National School of Engineering of Le Mans (ENSIM), France. She received the Diploma in telecommunications engineering, in 2006 and the M.Sc. degree in communications systems, in 2007 from the National Engineering School of Tunis (ENIT), Tunisia. She joined the Grenoble Automatic Signal-Speech-Images (GIPSA-Lab) as a Ph.D. Student in 2010. She received the Ph.D. degree in Signal, Image, Speech and Telecoms (SIPT) from the University of Grenoble, France, in 2012. Maha BEN ZID is the co-author of three book chapters and has also served as a reviewer for international conferences. Her research topics are multiple antenna systems and wireless sensor networks.

Contents

Preface

Multiple Input Multiple Output (MIMO) technology is a viable way to achieve high data rates for wireless communications. Previous researches have been done on the theoretical front and has been validated in practice. MIMO systems exploit both the diversity gain and the spatial multiplexing gain to allow for a substantial increase of spectral efficiency and higher data rates.

Recent focus is on extending the benefits of MIMO technology to multi-user systems. It is shown that Multi-User Multiple-Input Multiple-Output (MU-MIMO) systems achieve high data rates envisioned for 3G (3rd Generation) standards and other standards beyond 3G such as 3rd Generation Partnership Project Long Term Evolution (3GPP LTE) and LTE-Advanced.

Chapter 1 will give information about the fundamentals of multi-user MIMO communication. The chapter will consider the benefits of multiple antenna systems. MU-MIMO systems will be introduced and main topics including system configuration, fields of applications, capacity region and precoding techniques will be examined.

Chapter 2 will present a literature survey of the single-cell opportunistic beamforming (OBF) /random beamforming (RBF). Recent investigations on multi-cell RBF systems will be introduced and achievable rates of multi-cell RBF in finite-SNR regime will be reviewed. Different configurations including multi-cell MIMO-BC system, MISO/MIMO RBF schemes and single-cell OBF/RBF will be presented.

Chapter 3 will examine the LTE and LTE-Advanced framework. This chapter will consider the interference cancellation in multi-user scenario and low-complexity interference-aware receiver structure will be discussed.

Chapter 4 will be dedicated to multi-user interference suppression. Distributed antenna network (DAN) and conventional cellular system with centralized antennas at the BS (CAN system) will be considered. Single-carrier (SC) frequency domain adaptive antenna array (SC-FDAAA) for the uplink transmission will be performed for multi-user interference suppression and performances of both DAN SC-FDAAA and CAN SC-FDAAA will be evaluated.

Chapter 5 will consider solutions for the incorporation of multi-user capabilities in IEEE 802.11n/ac for WLAN systems. This chapter will be dedicated to the study of multi-user MAC layer protocols. Multi-user mechanism included in the 802.11ac draft standard and the additional challenges that arise from simultaneous transmissions will be examined.

Chapter 6 will survey the physical layer security for multi-user MIMO communications and suboptimal low-complexity transmission schemes will be discussed within challenging networks.

Chapter 7 will be dedicated to user selection based error probability of MIMO detector in multi-user MIMO system. Low complexity greedy user selection approach with suboptimal Lattice Reduction (LR)-based detector will be studied.

This book focuses on multi-user MIMO communication and covers a collection of the major topics and issues in multi-user MIMO systems. « Recent Trends in Multi-user MIMO Communications » provides a tutorial overview of the latest technologies and research keys related to multi-user communication.

The target audience of this book comprises researchers, Ph.D. students, academicians and professionals working in the field of Signal Processing and Multiple Antenna Systems. It provides an interdisciplinary guide that can help the reader about the multi-user MIMO communication.

Maha BEN ZID
National School of Engineering of Le Mans (ENSIM)
France

Section 1

Fundamentals of Multi-User MIMO Communication

Chapter 1

Multi User MIMO Communication: Basic Aspects, Benefits and Challenges

Ben Zid Maha and Raoof Kosai

Additional information is available at the end of the chapter

http://dx.doi.org/10.5772/57133

1. Introduction

The explosive growth of Multiple Input Multiple Output (MIMO) systems has permitted for high data rate and a wide variety of applications. Some of the technologies which rely on these systems are IEEE 802.11, Third Generation (3G) and Long Term Evolution (LTE) ones. Recent advances in wireless communication systems have contributed to the design of multi-user scenarios with MIMO communication. These communication systems are referred as multi-user MIMOs. Such systems are intended for the development of new generations of wireless mobile radio systems for future cellular radio standards. This chapter provides an insight into multi-user MIMO systems. We firstly present some of the main aspects of the MIMO communication. We introduce the basic concepts of MIMO communication as well as MIMO channel modeling. Thereafter, we evaluate the MIMO system performances. Then, we concentrate our analysis on the multi-user MIMO systems and we provide the reader a conceptual understanding with the multi-user MIMO technology. To do so, we present the communication system model for such emerging technology and we give some examples which describe the recent advances for multi-user MIMO systems. Finally, we introduce linear precoding techniques which could be exploited in multi-user MIMO systems in order to suppress inter-user interference.

2. MIMO communication

2.1. An historical overview

The main historical events which make the MIMO systems [2][3] are summarized as follows:

- In 1984, Jack Winters at Bell Laboratories wrote a patent on wireless communications using multiple antennas. Jack Winters in [4] presented a study of the fundamental limits on the data rate of multiple antenna systems in a Rayleigh fading environment.

- In 1993, Arogyaswami Paulraj and Thomas Kailath proposed the concept of spatial multiplexing using MIMO.
- Several articles which focused on MIMO concept were published in the period from 1986 to 1995 [5]. This was followed by the work of Greg Raleigh and Gerard Joseph Foschini in 1996 which invented new approaches involving space time coding techniques. These approaches were proved to increase the spectral efficiency of MIMO systems.
- In 1999, Thomas L. Marzetta and Bertrand M. Hochwald published an article [6] which provides a rigorous study on the MIMO Rayleigh fading link taking into consideration information theory aspects.
- The first commercial MIMO system was developed in 2001 by Iospan Wireless Inc.
- Since 2006, several companies such as Broadcom and Intel have introduced a novel communication technique based on the MIMO technology for improving the performance of wireless Local Area Network (LAN) systems. The new standard of wireless LAN systems is named IEEE 802.11n.

Nowadays, MIMO systems are implemented in many advanced technologies such as various standard proposals for the Fourth Generation (4G) of wireless communication systems and LTE. MIMO technology was shown to boost the communication system capacity and to enhance the reliability of the communication link since it uses several diversity schemes beyond the spatial diversity.

2.2. Fundamentals of MIMO system

MIMO system model is depicted in Figure 1. We present a communication system with N_T transmit antennas and N_R receive antennas.

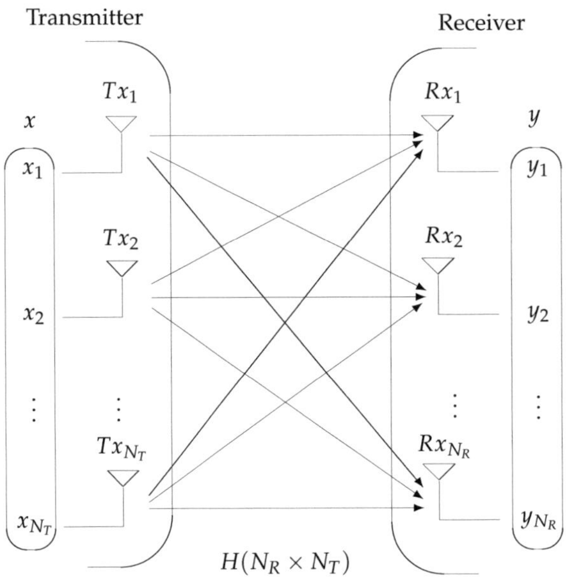

Figure 1. MIMO system model

http://dx.doi.org/10.5772/57133

Antennas $Tx_1, \ldots, Tx_{N_T}$ respectively send signals $x_1, \ldots, x_{N_T}$ to receive antennas $Rx_1, \ldots, Rx_{N_R}$. Each receive antenna combines the incoming signals which coherently add up. The received signals at antennas $Rx_1, , \ldots, Rx_{N_R}$ are respectively denoted by $y_1, \ldots, y_{N_R}$. We express the received signal at antenna Tx_q; $q = 1, \ldots, N_R$ as:

$$y_q = \sum_{p=1}^{N_T} h_{qp} \cdot x_p + b_q \quad ; q = 1, \ldots, N_R \tag{1}$$

The flat fading MIMO channel model is described by the input-output relationship as:

$$y = H \cdot x + b \tag{2}$$

- H is the $(N_R \times N_T)$ complex channel matrix given by:

$$H = \begin{pmatrix} h_{11} & h_{12} & \ldots & h_{1N_T} \\ h_{21} & h_{22} & \ldots & h_{2N_T} \\ \vdots & \vdots & \ddots & \vdots \\ h_{N_R1} & h_{N_R2} & \ldots & h_{N_RN_T} \end{pmatrix}$$

  h_{qp}; $p = 1, \ldots, N_T$; $q = 1, \ldots, N_R$ is the complex channel gain which links transmit antenna Tx_p to receive antenna Rx_q.
- $x = [x_1, \ldots, x_{N_T}]^T$ is the $(N_T \times 1)$ complex transmitted signal vector.
- $y = [y_1, \ldots, y_{N_R}]^T$ is the $(N_R \times 1)$ complex received signal vector.
- $b = [b_1, \ldots, b_{N_R}]^T$ is the $(N_R \times 1)$ complex additive noise signal vector.

The continuous time delay MIMO channel model of the $(N_R \times N_T)$ MIMO channel H associated with time delay τ and noise signal $b(t)$ is expressed as:

$$y(t) = \int_{\tau} H(t, \tau)\, x(t - \tau)\, d\tau + b(t) \tag{3}$$

- $y(t)$ is the spatio-temporel output signal.
- $x(t)$ is the spatio-temporel input signal.
- $b(t)$ is the spatio-temporel noise signal.

$(\cdot)^T$ denotes the transpose operator.

2.3. MIMO channel modeling

Several MIMO channel models [7] have been proposed in literature. These models mainly fall into two categories as depicted in Figure 2.

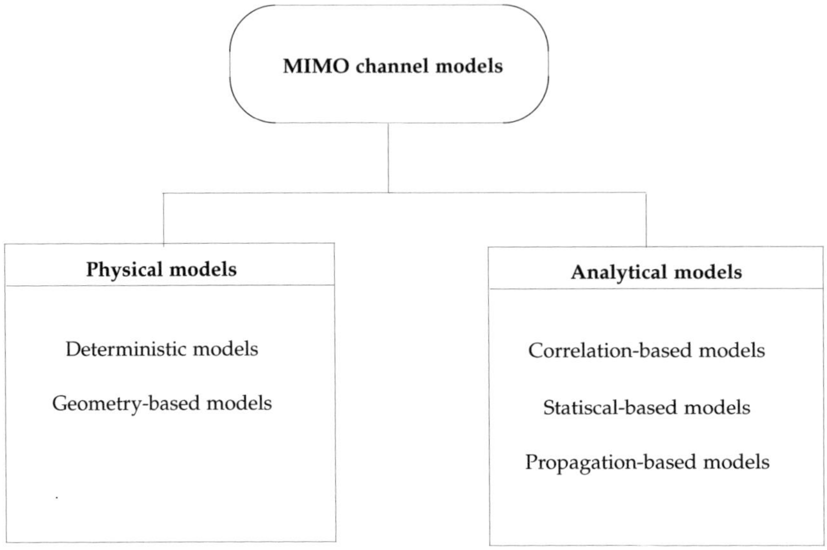

Figure 2. MIMO channel propagation models

2.3.1. *Physical models*

MIMO channel impulse response is evaluated according to the radio wave which propagates from the transmitter to the receiver. The MIMO channel model is determined based on the experimental measurements made for extracting channel propagation parameters including antenna configuration at both the transmitter and the receiver, antenna polarization, scatterers,... Physical models include both deterministic models and Geometry-based stochastic channel models (GSCMs).

- Deterministic models define a channel model according to the prediction of the propagation signal.
- Geometry-based Stochastic Channel Models (GSCMs) have an immediate relation with the physical characteristics of the propagation channel. These models suppose that clusters of scatterers are distributed around the transmitter and the receiver. The scatterers locations are defined according to a random fashion that follows a particular probability distribution. Scatterers result in discrete channel paths and can involve statistical characterizations of several propagation parameters such as delay spread, angular spread, spatial correlation and cross polarization discrimination. We distinguish two possible schemes which are the Double Bounce Geometry-based Stochastic Channel Models (DB-GSCMs) and the Single Bounce Geometry-based Stochastic Channel Models (SB-GSCMs). That is when a single bounce of scatterers is placed around the transmit antennas or receive antennas.

2.3.2. *Analytical models*

The second class of MIMO channel models includes analytical models which are based on the statistical properties obtained through measurement (Distribution of Direction of Departure (DOD), distribution of Direction of Arrival (DOA),...). Analytical channel models can be

classified into correlation-based models (such as i.i.d model, Kronecker model, Keyhole model,...), statistical -based models (such as Saleh-Valenzuela model and Zwick model) and propagation-based models (such as Müller model and Finite scatterer model).
We provide in [8] a detailed description of MIMO systems with geometric wide MIMO channel models where advanced polarization techniques [9][10] are exploited.

2.4. MIMO system performances

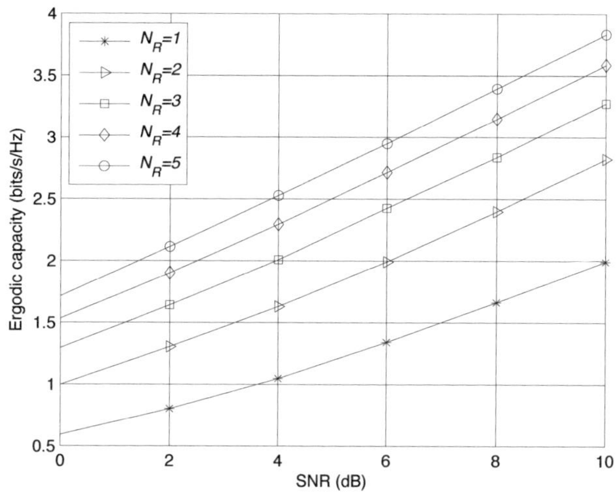

Figure 3. Ergodic capacity for MIMO systems

MIMO technology has been shown to improve the capacity of the communication link without the need to increase the transmission power. MIMO system capacity is mainly evaluated according to the following scenarios:

1. When no Channel State Information (CSI) is available at the transmitter, the power is equally split between the N_T transmit antennas, the instantaneous channel capacity is then given by:

$$C(H) = \log_2 \left[\det \left(I_{N_R} + \frac{\gamma}{N_T} \cdot HH^* \right) \right] \quad bits/s/Hz \tag{4}$$

γ denotes the Signal to Noise Ratio (SNR).
$(\cdot)^*$ stands for the conjugate transpose operator.

2. When CSI is available at the receiver, Singular Value Decomposition (SVD) is used to derive the MIMO channel capacity which is given by:

$$C_{SVD}(H) = R \cdot \log_2 \left[\det \left(1 + \frac{\gamma}{N_T} HH^* \right) \right] \quad bits/s/Hz \tag{5}$$

$R = min(N_R, N_T)$ is the rank of the channel matrix H

3. When CSI is available at both the transmitter and the receiver, the channel capacity is computed by performing the water-filling algorithm. The instantaneous channel capacity is then:

$$C_{WF}(H) = \sum_{p=1}^{R} \log_2 \left[\left(\frac{\lambda_{H,p} \cdot \mu}{\sigma_b^2} \right)^+ \right] \quad bits/s/Hz \tag{6}$$

- $a^+ = \max(a, 0)$
- $\lambda_{H,p}$ is the $p-th$ singular value of the channel matrix H
- μ is a constant scalar which satisfies the total power constraint
- σ_b^2 is the noise signal power

We consider the case where CSI is available at the receiver, the simulated ergodic MIMO capacity is depicted in Figure 3. For a MIMO system with two transmit antennas, numerical results show that ergodic capacity linearly increases with the number of antennas. Plotted curves are presented for different levels of the SNR. The use of additional antennas improves the performances of the communication system. Moreover, MIMO system takes advantage of multipath propagation. The performances of MIMO system are observed in the following in terms of the Bit Error Rate (BER). We consider a MIMO system with various receive antennas, the BER is evaluated for communication systems with Rayleigh fading MIMO channel and additive gaussian noise. At the receive side, the Maximum Ratio Combining (MRC) technique is performed. According to Figure 4, it is obvious that MIMO technology allows for a significant improvement of the BER.
Once the MIMO technology is presented, we introduce in the following multi-user MIMO systems.

3. Multi-user MIMO system

The growth in MIMO technology has led to the emergence of new communication systems. We are particularly interested in this chapter in multi-user MIMO (MU-MIMO) ones [11]. MU-MIMO [12] system is often considered in literature as an extension of Space-Division Multiple Access (SDMA). This technology supports multiple connections on a single conventional channel where different users are identified by spatial signatures. SDMA uses spatial multiplexing and enables for higher data rate. This could be achieved by using multiple paths as different channels for carrying data. Another benefit of using the SDMA technique in cellular networks is to mitigate the effect of interference coming from adjacent cells. Traditional communication MIMO systems are usually referred as single-user MIMO systems (SU-MIMOs) or also point-to-point MIMO. Case of MIMO systems, the access point communicates with only one mobile terminal (the user). Both the access point and the mobile terminal are equipped with multiple antennas. In contrast to the single-user case, the access point is able to communicate with several mobile terminals. SU-MIMO and MU-MIMO systems are two possible configurations for multi-user communication systems. We also find other configurations in literature such as MU-MIMO with cooperation where cooperation is

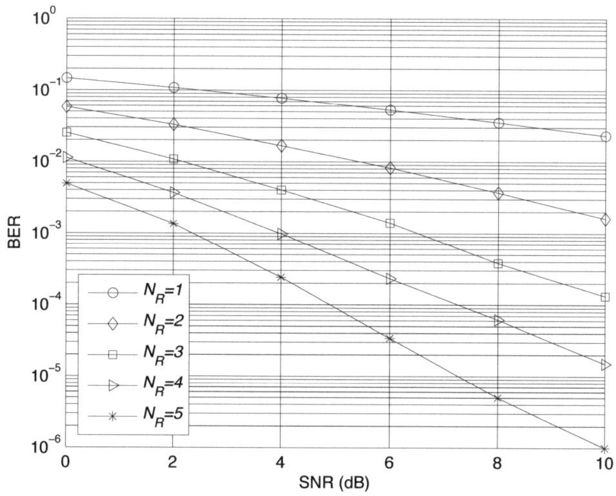

Figure 4. Improvement of the BER for MIMO ($N_R \times 2$) as a function of receive antennas number

established between base stations [2]. Basic configurations of downlink multi-user MIMO systems are depicted in Figure 5. Figure 5(a) represents the SU-MIMO system where a Base Station (BS) equipped with antennas $Tx_1, \ldots, Tx_N$ communicates with user U which is equipped with M antennas $Rx_1, \ldots, Rx_M$. In Figure 5(b), the presented MU-MIMO system consists of two base stations BS_1 and BS_2 each one is equipped with N antennas. Generalized MU-MIMO systems may consist of more base stations where the number of antennas could be different. At the receive side, K users $U_1, \ldots, U_K$ with respectively $M_1, \ldots, M_K$ antennas communicate with the transmit base stations. The same communication model is performed for the MU-MIMO with cooperation (Figure 5(c)) where cooperation is established between BS_1 and BS_2.
Once multi-user communication systems are introduced, we explain in the following section the difference between SU-MIMO and MU-MIMO configurations.

4. MU-MIMO vs SU-MIMO

Table 1 summarizes the main features of both SU-MIMO and MU-MIMO systems [13]. In contrast to MU-MIMO systems where one base station could communicate with multiple users, base station only communicate with a single user in the case of SU-MIMO systems. In addition, MU-MIMO systems are intended to employ multiple receivers so that to improve the rate of communication while keeping the same level of reliability. These systems are able to achieve the overall multiplexing gain obtained as the minimum value between the number of antennas at base stations and the number of antennas at users. The fact that multiple users could simultaneously communicate over the same spectrum improves the system performance. Nevertheless, MU-MIMO networks are exposed to strong co-channel interference which is not the case for SU-MIMO ones. In order to solve the problem of interference in MU-MIMO systems, several approaches have been proposed for interference management [14][15]. Some of these approaches are based on beamforming technique [31]. Moreover, in contrast to SU-MIMO systems, MU-MIMO systems require perfect CSI in

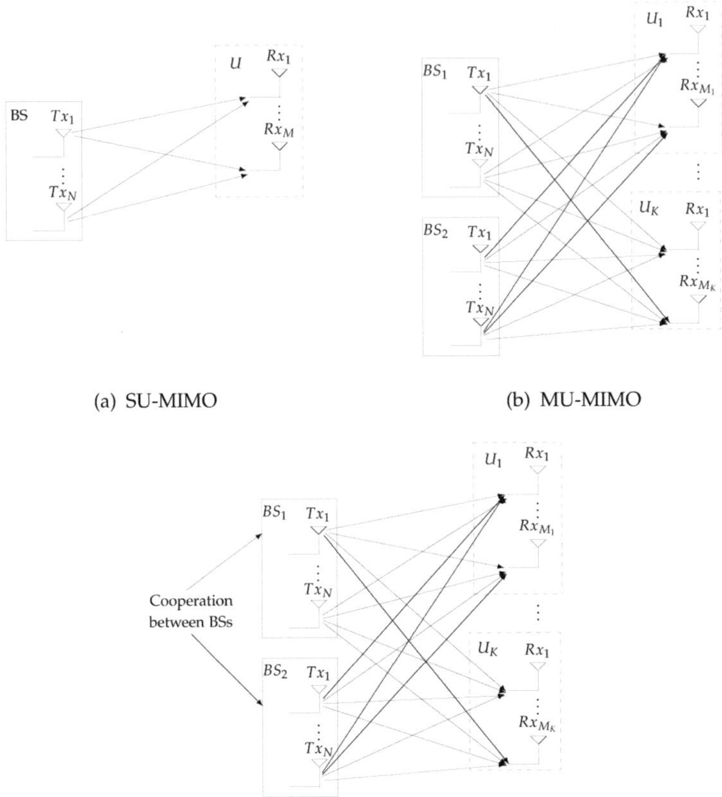

(a) SU-MIMO

(b) MU-MIMO

(c) MU-MIMO with cooperation

Figure 5. MU-MIMO configurations

order to achieve high throughput and to improve the multiplexing gain [16]. Finally, the performances of MU-MIMO and SU-MIMO systems in terms of throughput depend on the SNR level. In fact, at low SNRs, SU-MIMO performs better. However, at high SNRs level, MU-MIMO provides better performances.

Feature	**MU-MIMO**	**SU-MIMO**
Main aspect	BS communicates with multiple users	BS communicates with a single user
Purpose	MIMO capacity gain	Data rate increasing for single user
Advantage	Multiplexing gain	No interference
CSI	Perfect CSI is required	No CSI
Throughput	Higher throughput at high SNR	Higher throughput at low SNR

Table 1. Comparison between SU-MIMO and MU-MIMO systems

5. Communication schemes for MU-MIMO systems

Communication schemes for MU-MIMO systems include both uplink MU-MIMO (UL-MU-MIMO) and downlink MU-MIMO (DL-MU-MIMO). Case of uplink communication,

users transmit signals to the base station. However, in the case of downlink communication, base station transmits signals to users. A representation of these systems is depicted in Figure 6. We assume that the base station is equipped with N antennas. Case of DL-MU-MIMO, the base station attempts to transmit signals to K users $U_1, \ldots, U_K$ which are respectively equipped with antennas of numbers $M_1, \ldots, M_K$.
For notations, if antenna k acts like a receiving antenna, it is denoted by Rx_k. Otherwise, it is denoted by Tx_k.

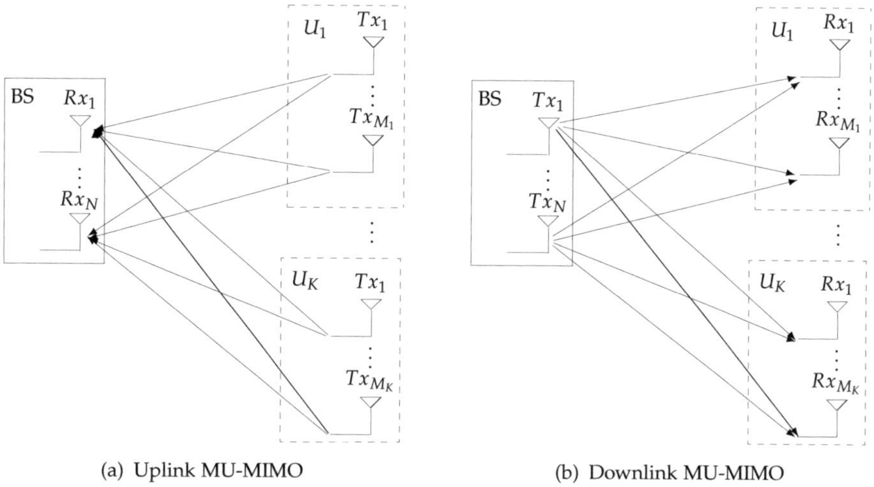

Figure 6. MU-MIMO communication models: UL-MU-MIMO and DL-MU-MIMO

5.1. UL-MU-MIMO

Let $X_k(M_k \times 1)$, the transmit signal vector of user U_k; $k = 1, \ldots, K$. We assume that data streams associated to user U_k; $k = 1, \ldots, K$ are zero mean white random vectors where :

$$\mathrm{E}\{X_k X_k{}^*\} = I_{M_k}; \quad k = 1, \ldots, K \tag{7}$$

E denotes the expected value operator.
The complex channel matrix relating user U_k; $k = 1, \ldots, K$ to the base station, H_k is of dimension $(N \times M_k)$. In presence of additive noise signal $b(N \times 1)$, the received signal vector at the base station, $y(N \times 1)$ is expressed in the slow fading model by:

$$y = \sum_{k=1}^{K} H_k \cdot X_k + b \tag{8}$$

The noise signal vector is a zero mean white Gaussian variable with variance σ_b^2. The uplink scenario should satisfy two constraints:

- It should be as many receive antennas at the base station as the total number of users antennas.

- Each user should have as many transmit antennas as the number of data streams.

In Figure 7, the block diagram for the UL-MU-MIMO includes a joint linear precoder and decoder. Linear precoders associated to users $U_1, \ldots, U_K$ will be respectively denoted by

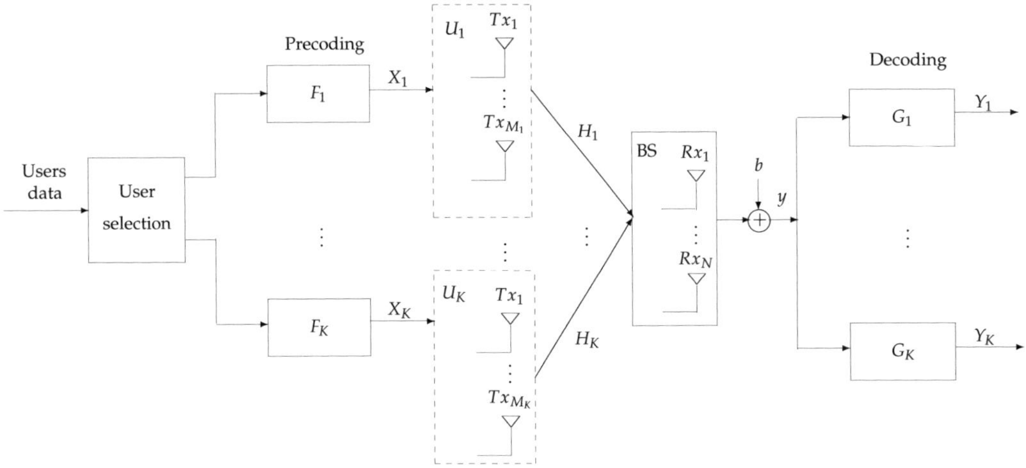

Figure 7. Block diagram for the UL-MU-MIMO with coding techniques: N antenna BS and K multiple antenna users

$F_1, \ldots, F_K$. The received signal vector at the BS is then expressed as :

$$y = \sum_{k=1}^{K} H_k \cdot F_k \cdot X_k + b \tag{9}$$

An estimate of the transmitted signal vectors denoted by $Y_k;\ k = 1, \ldots, K$ are obtained by using the linear decoders $G_1, \ldots, G_K$. The decoding process is such that :

$$Y_k = G_k \cdot y$$

5.2. DL-MU-MIMO

DL-MU-MIMO communication model assumes that K users are simultaneously receiving signals from the base station. The transmitted signal vector $x(N \times 1)$ is expressed as the sum of signals intended to users $U_1, \ldots, U_K$:

$$x = \sum_{k=1}^{K} X_k \tag{10}$$

The channel matrix between user $U_k;\ k = 1, \ldots, K$ and the base station is denoted by $H_k(M_k \times N)$. At each user, received signal vector of dimension $(M_k \times 1);\ k = 1, \ldots, K$ is given by:

$$Y_k = H_k \cdot x + B_k \quad ; \ k = 1, \ldots, K \tag{11}$$

http://dx.doi.org/10.5772/57133

$B_k; k = 1, \ldots, K$ is an additive noise signal vector of size $(M_k \times 1)$. Equation (11) could be also written:

$$Y_k = H_k \cdot x + B_k \tag{12}$$

$$= H_k \cdot X_k + \sum_{j \neq k}^{K} (H_k \cdot X_j) + B_k \quad ; \; k = 1, \ldots, K \tag{13}$$

The second term of the sum in equation (13) represents the interference signal coming from multiple users. Processing techniques such as beamforming should be introduced in the block diagram of the MU-MIMO system for mitigating the effect of users interference and improving the performances of the communication system.

6. Fields of application

MU-MIMO technology finds its applications in many areas and is nowadays exploited in many evolving technologies wich are described in the following.

3GPP LTE: 3rd Generation Partnership Project (3GPP) Long Term Evolution (LTE) is one of the next generation cellular networks which exploit the MU-MIMO technology. Thanks to this technology, available radio spectrum 3GPP LTE networks could achieve higher spectral efficiencies than existing 3G networks [18][19].

Release 8 of LTE: The first release of LTE (Release 8) was commercially deployed in 2009. Release 8 has introduced SU-MIMO scheme in the communication system model. This release only uses one transmission mode (Transmission mode 5) which has been defined for MU-MIMO systems. Transmission mode 5 supports rank 1 transmission for two User Equipments (UEs). In order to achieve the performances of MU-MIMO systems, feedback parameters such as the channel Rank Indicator (RI) and the Channel Quality Indicator (CQI)/Precoding Matrix Indicator (PMI) feedback [17] are required.

Release 9 of LTE: The second release of LTE (Release 9) provides enhancements to Release 8. The LTE Release 9 supports transmission mode 8 and includes both SU-MIMO and MU-MIMO schemes.

LTE advanced: Other progress in LTE MIMO systems have been obtained through LTE advanced. The performed mode is the transmission mode 9. This mode allows for a possible switch between SU-MIMO and MU-MIMO.

Multiple-cell networks: MU-MIMO systems have received wide spread success in wireless networks. Examples of applications include the multiple-cell networks [21] with multiple access channels where possible coordination among base stations is established. Figure 8 shows a MU-MIMO coordinated network in a cellular network. Three classes of cells are presented. These cells are referred as :

- Coordinated cells
- Central cell
- Interfering cells

The coordination between cells is performed by the Central Station (CS). The aim of this coordination is to mitigate the effect of inter-cells interference. Coding techniques should be employed in order to mitigate the effect of interfering cells.

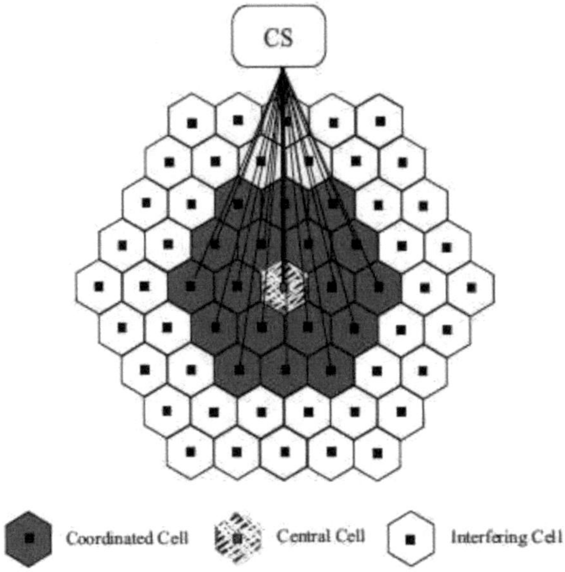

Figure 8. MU-MIMO coordinated network in cellular network [20]

Digital Subscriber Line (DSL): MU-MIMO systems are not only performed by multi-cell systems but also find their applications in other systems such as the downlink of a Digital Subscriber Line (DSL)[22][13].

The performance of MU-MIMO could be improved via the use of Orthogonal Frequency Division Multiplexing (OFDM) or Orthogonal Frequency Division Multiplexing Access (OFDMA) for multiple access scenarios in frequency selective channels. MU-MIMO systems could also improve multi-user diversity by performing High Data Rate (HDR) or Code Division Multiple Access (CDMA) techniques.

7. Capacity region of multi-user MIMO system

There is no closed form for the channel capacity of multi-user MIMO systems. For this purpose, the performances of such systems will be analyzed in terms of the capacity region. This metric [23] could be defined in the usual Shannon sense as the highest rates that can be achieved with arbitrarily small error probability. Firstly, the capacity [24] needs to be evaluated for each user. Then, the capacity region is determined as the entire region for which maximum achievable rates are reached. The evaluation of the capacity region is strongly related to some constraints and should be determined according to the performed communication scenario.
We address the following scenarios :

1. UL-MU-MIMO with single antenna users
2. UL-MU-MIMO with multiple antenna users
3. DL-MU-MIMO with multiple antenna users and single antenna BS

http://dx.doi.org/10.5772/57133

7.1. Capacity region of UL-MU-MIMO with single antenna users

We consider the UL-MU-MIMO with N multiple antenna BS and K single antenna users. The performed communication scheme is depicted in Figure 9. The equivalent MIMO channel for the presented communication model is given by:

$$H = [H_1, \ldots, H_K] \tag{14}$$

$H_k(N \times 1);\ k = 1, \ldots, K$ represents the Single Input Multiple Output (SIMO) channel between user $U_k;\ k = 1, \ldots, K$ and the BS. Case of two users (i.e. K=2), the capacity region is defined as the set of rates (R_1, R_2) associated to users U_1 and U_2.
We consider the notations:

- P_1: average power constraint on user U_1
- P_2: average power constraint on user U_2
- N_0: noise signal power

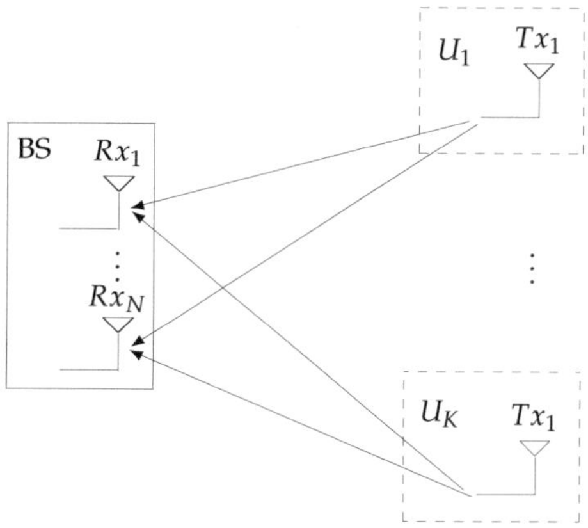

Figure 9. UL-MU-MIMO with N multiple antenna BS and K single antenna users

The capacity region [25] is evaluated by determining the individual rate constraint for each user. Assuming that user U_1 has the entire channel, an upper bound of the maximum achievable rate is given by :

$$R_1 \leq \log_2\left(1 + \frac{\|H_1\|^2 \cdot P_1}{N_0}\right) \tag{15}$$

$\|\cdot\|$ indicates the Frobenius norm.
Similarly, an upper bound for the maximum achievable rate for user U_2 is:

$$R_2 \leq \log_2\left(1 + \frac{\|H_2\|^2 \cdot P_2}{N_0}\right) \tag{16}$$

Finally, the sum rate constraint which is obtained when both users are acting as two transmit antennas of a single user has an upper bound expressed as :

$$R_1 + R_2 \leq \log_2 \left[\det \left(I_N + \frac{H.diag(P_1, P_2).H^*}{N_0} \right) \right] \tag{17}$$

The capacity region for the UL-MU-MIMO is presented in Figure 10 where two users with single antennas are considered.

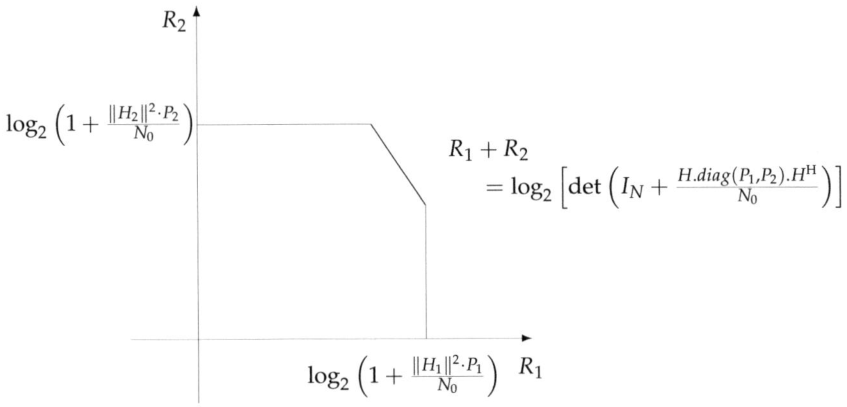

Figure 10. Capacity region of UL-MU-MIMO for two single antenna users

Case of K users, the capacity region is determined as a function of several constraints and $K!$ corner points are determined for evaluating the boundary of the capacity region. For rates $R_1, \ldots, R_K$ respectively associated to users $U_1, \ldots, U_K$, the sum rate is determined for an optimal receiver [25] as:

$$\sum_{k \in S} R_k \leq \log_2 \left[\det \left(I_N + \frac{1}{N_0} \sum_{k \in S} P_k \cdot H_k \cdot H_k^* \right) \right]; \quad S \subset \{1, \ldots, K\} \tag{18}$$

7.2. Capacity region of UL-MU-MIMO with multiple antenna users

The capacity region could be obtained for the generalized case where the base station has N antennas and user U_k; $k = 1, \ldots, K$ is equipped with multiple antennas of number $M_k > 1$. An upper bound of the maximum achievable rate for user U_k is given by :

$$R_k \leq \log_2 \left[\det \left(I_N + \frac{H_k \cdot D_k \cdot H_k^*}{N_0} \right) \right] \quad ; k = 1, \ldots, K \tag{19}$$

- $H_k(N \times M_k)$ links the N antenna base station to the M_k antenna user; $k = 1, \ldots, K$.
- $D_k(M_k \times M_k)$ is a diagonal matrix formed by the power allocated at transmit antennas at user U_k.

The sum rate constraint of UL-MU-MIMO with multiple antennas users is expressed as :

$$R_1 + \ldots + R_K \leq \log_2 \left[\det \left(I_N + \frac{\sum_{k=1}^{K} H_k.D_k.H_k^*}{N_0} \right) \right] \quad (20)$$

7.3. DL-MU-MIMO with multiple antenna users and single antenna BS

In the case of downlink scenario, the upper bounds of the users rates are analogously determined as the uplink scenario. Nevertheless, the effect of interference could not be neglected. In fact, for the scenario with two multiple antenna users U_1 and U_2 and one antenna base station, the upper bounds of the rates achievable by users U_1 and U_2 become:

$$R_1 \leq \log_2 \left(1 + \frac{\|H_1\|^2 \cdot P_1}{N_0 + \|H_1\|^2 \cdot P_2} \right) \quad (21)$$

and

$$R_2 \leq \log_2 \left(1 + \frac{\|H_2\|^2 \cdot P_2}{N_0} \right) \quad (22)$$

Here, the signal of user U_2 is considered as interference for user U_1.

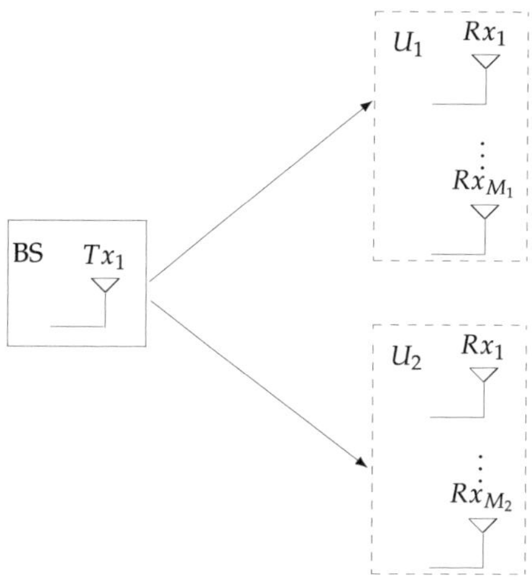

Figure 11. DL-MU-MIMO with multiple antenna users and single antenna BS

8. Precoding techniques

The DL-MU-MIMO system uses precoding techniques which are usually linear.

8.1. Zero Forcing and Block Diagonalization methods

Popular low-complexity techniques include both Zero Forcing (ZF) and Block Diagonalization (BD)[27][28] methods. Algorithms for the ZF as well as BD methods are presented in [29]. The aim of these solutions is to improve the sum rate capacity of the communication system under a given power constraint. These performances could be achieved by canceling inter-user interference. Zero Forcing Dirty Paper Coding (DPC) [30] represents a famous technique for data precoding where the channel is subject to interference which is assumed to be known at the transmitter. The precoding matrix is equal to the conjugate transpose of the upper triangular matrix obtained via the QR decomposition of the channel matrix.

8.1.1. MU-MIMO with Block Diagonalization precoding

We consider a communication system model with a broadcast MIMO channel where the transmitter is a base station equipped with N antennas and the receiver consists of K users $U_k;\ k = 1 \dots K$ (See figure 6(b)). The received signal at user $U_k;\ k = 1 \dots K$ with dimension $(M_k \times 1)$ is expressed as :

$$Y_k = H_k \cdot {V_{BD}}^{(k)} \cdot X_k + B_k \qquad ;\ k = 1, \dots, K \tag{23}$$

- $H_k(M_k \times N)$ is the channel matrix between user U_k and the base station
- ${V_{BD}}^{(k)}(N \times M_k)$ is the BD precoding matrix for user U_k
- X_k is the transmit signal for user U_k
- $B_k(M_k \times 1)$ is the additive noise signal vector

We assume in the following that users $U_1, \dots U_K$ have the same number of antennas which will be denoted by M. Block Diagonalization strategy defines a set of precoding matrices ${V_{BD}}^{(k)}(N \times M)$ associated to users $U_1, \dots, U_K$. These matrices form an orthonormal basis such that:

$$[{V_{BD}}^{(k)}]^* \cdot {V_{BD}}^{(k)} = I_M \quad ;\ k = 1 \dots K \tag{24}$$

and the Block Diagonalization algorithm achieves :

$$H_k \cdot {V_{BD}}^{(j)} = 0 \quad ;\ \forall\ j \neq K \tag{25}$$

The aim of these conditions is to eliminate multi-user interference so that to maximize the achievable throughput.
The performance of downlink communication scenarios with precoding techniques depends on the SNR level. In fact, it has been shown in [27] that SU-MIMO achieves better performances than MU-MIMO at low SNRs. However, the BD MU-MIMO achieves better performances at high SNRs. As such, switching between SU-MIMO and MU-MIMO is optimal for obtaining better total rates over users.

http://dx.doi.org/10.5772/57133

8.1.2. MU-MIMO with Zero Forcing precoding

Case of Zero Forcing strategy, each transmitted symbol to the $l - th$ antenna (among M antennas of user U_k) is precoded by a vector which is orthogonal to the columns of H_j, $j \neq k$ but not orthogonal to the $l - th$ column of H_k [26].

8.2. Beamforming for linear precoding

Beamforming paradigms represent another class of linear precoding for MU-MIMO systems. For the communication model with beamforming (Figure 12), we consider a MU-MIMO system with K multiple antenna users $U_1, \ldots, U_K$ at the receive side which are respectively equipped with $M_1, \ldots, M_K$ antennas. At the transmit side, a multiple antenna base station with N antennas transmits data signals $x_1, \ldots, x_K$ to users $U_1, \ldots, U_K$.

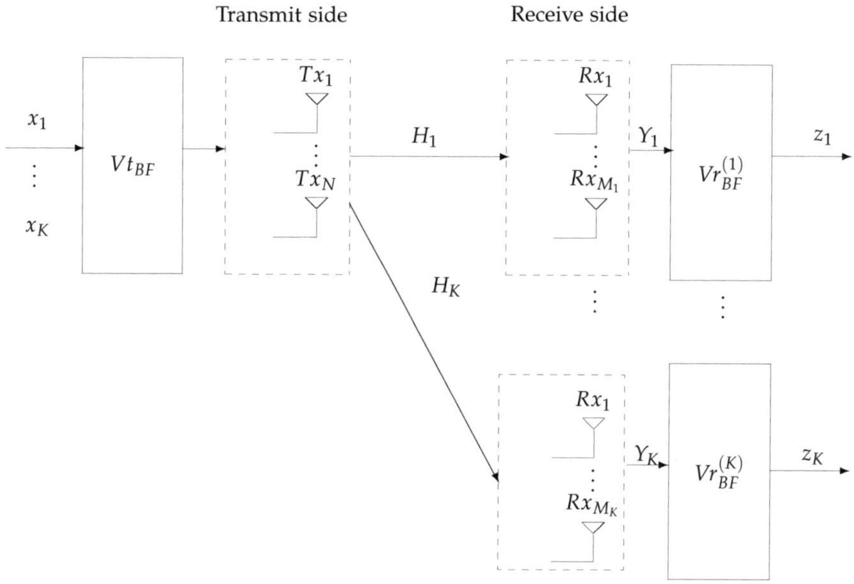

Figure 12. MU-MIMO with beamforming

The received signal vector at user U_k; $k = 1, \ldots, K$ is expressed as :

$$Y_k = H_k \cdot Vt_{BF}^{(k)} \cdot x_k + \sum_{j=1,\ k \neq j}^{K} H_k \cdot Vt_{BF}^{(j)} \cdot x_j + B_k \quad ; \ k = 1, \ldots, K \tag{26}$$

where:

- $H_k(M_k \times N)$ is the complex channel matrix between receiver U_k and the transmit base station.
- $B_k(M_k \times 1)$ is an additive noise signal vector.

- ${Vt_{BF}}^{(k)}(N \times 1)$ is the transmit beamforming vector of index k. The transmit beamforming matrix is :

$$Vt_{BF} = [Vt_{BF}^{(1)}, \ldots, Vt_{BF}^{(k)}] \tag{27}$$

At the receive side, beamforming vectors are denoted by $Vr_{BF}^{(k)}(M_k \times 1)$.

$$Vr_{BF}^{(k)} = [vr_{BF}^{(1)}, \ldots, vr_{BF}^{(M_k)}]^T \tag{28}$$

The resulting signal at user U_k is:

$$z_k = {Y_k}^* \cdot Vr_{BF}^{(k)} \quad ; \; k = 1, \ldots, K \tag{29}$$

The conjoint receive-transmit beamforming weights are obtained by maximizing the sum rate of the MU-MIMO system expressed as:

$$R_{sum} = \sum_{k=1}^{K} \log_2(1 + SINR^{(k)}) \tag{30}$$

$SINR^{(k)}$; $k = 1, \ldots, K$ is the Single Interference Noise Ratio (SINR) [31] associated to user U_k. The SINR is determined as the ratio of the received strength for the desired signal to the strength of undesired signal obtained as the sum of noise and interference signal.
For unit signal noise variance, the SINR for user U_k is given by :

$$SINR^{(k)} = \frac{\| Vr_{BF}^{(k)} \cdot H_k^* \cdot Vt_{BF}^{(k)} \|^2}{\left(\sum_{k=1}^{K} \| Vr_{BF}^{(k)} \cdot H_k^* \cdot Vt_{BF}^{(k)} \|^2 \right) + 1} \tag{31}$$

Beamforming weights at the receiver are determined so that to suppress inter-user interference such as [32]:

$$Vr_{BF}^{(k)} = \frac{[C^{(k)}]^{-1} \cdot H_k \cdot Vr_{BF}^{(k)}}{\| [C^{(k)}]^{-1} \cdot H_k \cdot Vt_{BF}^{(k)} \|_2} \tag{32}$$

$C^{(k)}$ is the covariance matrix of H_k.
$\| \cdot \|_2$ stands for the 2-norm operator.

9. Conclusion

This chapter presents a basic introduction to the fundamentals of multi-user MIMO communication. MU-MIMO is considered as an enhanced form of MIMO technology. Such technology has been a topic of extensive research since the last three decades. The attractive features of MIMO systems have shown that the use of multiple antennas at both the ends of the communication link significantly improves the spectral efficiency of the communication system as well as the reliability of the communication link.

In multiuser channels and cellular systems, MIMO is offered for MU-MIMO communication to allow for spatial sharing of the channel by several users.

Nowadays, it has been a great deal with MU-MIMO systems. Several approaches are adopted and different scenarios may be considered. Throughout this chapter, we have presented possible configurations associated with MU-MIMO with particular emphasis on the fundamental differences between SU-MIMO and MU-MIMO.

Some scenarios have been considered for performance evaluation of MU-MIMO communication in terms of the capacity region metric.

MU-MIMO scenarios follow into UL-MU-MIMO for Multiple Access Channel (MAC) and the DL-MU-MIMO for Broadcast Channel (BC). The DL-MU-MIMO is the more challenging scenario since optimum strategies for interference cancelation are required.

Througout this chapter, we have presented precoding techniques used within MU-MIMO systems for efficient transmission and interference cancelation. Among the existing techniques, we have introduced ZF and BD methods. Of particular interest, we have described the linear beamforming algorithms.

The design of multi-user MIMO systems is attractive for the research field as well for the industrial one and the field of application is extensively growing.

Author details

Ben Zid Maha[1,*] and Raoof Kosai[1,2]

* Address all correspondence to: maha.ben-zid@gipsa-lab.grenoble-inp.fr

1 École Nationale Supérieure d'Ingénieurs du Mans (ENSIM), Le Mans, France
2 Laboratoire d'Acoustique de l'Université du Maine, (LAUM), UMR CNRS 6613, Le Mans, France

References

[1] Ben Zid, M., Raoof, K. and Bouallègue, A. (2012). MIMO spectral efficiency over energy consumption requirements: Application to WSNs, *Int'l J. of Communications, Network and System Sciences* Vol. 5(No. 2): 121–129.

[2] Ben Zid, M., Raoof, K. and Bouallègue, A. (2012). MIMO Systems and Cooperative Networks Performances, *in: Advanced Cognitive Radio Network*, Scientific Research Publishing, ISBN: 193–506–874–1.

[3] Raoof, K. and Zhou, H. (2009). *Advanced MIMO systems*, Scientific Research Publishing, ISBN:978–1–935068–02–0.

[4] Winters, J.H. (1987). On the capacity of radio communication systems with diversity in a Rayleigh fading environment, *IEEE Journal on Selected Areas in Communications* Vol. 5(No. 5): 871–878.

[5] Telatar, I. E.(1995). Capacity of multi-antenna Gaussian channels, *European Transactions on Telecommunications* Vol. 10(No. 6): 585–595.

[6] Marzetta, T. L. and Hochwald, B. M. (1999). Capacity of a mobile multiple-antenna communication link in Rayleigh flat fading, *IEEE Transactions on Information Theory* Vol. 45(No. 1): 139–157.

[7] Almers, P., Bonek, E., Burr, A., Czink, N.and Debbah, M.(2007). Survey of channel and radio propagation models for wireless MIMO systems, *EURASIP Journal on Wireless Communications and Networking* Vol. 2007(No. 1): 1–19.

[8] Ben Zid, M.(2012). Emploi de techniques de traitement de signal MIMO pour des applications dédiées réseaux de capteurs sans fil,*Thesis dissertation-UJF-Grenoble I.* http://tel.archives-ouvertes.fr/tel-00745006.

[9] Prayongpun, N. (2009). Modélisation et étude de la capacité du canal pour un système multi-antennes avancé exploitant la diversité de polarisation, *Thesis dissertation-UJF-Grenoble I* http://tel.archives-ouvertes.fr/tel-00396666/en/

[10] Dao, M.T., Nguyen, V.A., Im, Y.T., Park, S.O. & Yoon,G.(2011). 3D Polarized channel modeling and performance comparison of MIMO antenna configurations with different polarizations, *IEEE Transactions on Antennas and Propagation* Vol. 59(No. 7): 2672–2682.

[11] Soysal, A. and Ulukus, S. (2007). Optimum power Allocation for Single-User MIMO and Multi-User MIMO-MAC with Partial CSI, *IEEE Transactions on Selected Areas in Communications* Vol. 25(No. 7): 1402–1412.

[12] Serbetli, S. and Yener, A. (2004). Transceiver Optimization for Multiuser MIMO Systems, *IEEE Transactions on Signal Processing* Vol. 52(No. 1): 214–226.

[13] Bengtsson, M. (2004). From single link MIMO to multi-user MIMO, *Proceedings of* IEEE International Conference on Acoustics, Speech, and Signal Processing (ICASSP '04), pp. iv-697–iv-700.

[14] Codreanu, M., Tolli, A., Juntti, M. & Latva-aho, M.(2007). Joint Design of Tx-Rx Beamformers in MIMO Downlink Channel, *IEEE Transactions on Signal Processing* Vol. 55(No. 9): 4639–4655.

[15] Gomadam, K., Cadambe, V. R., & Jafar, S. A.(2011). A Distributed Numerical Approach to Interference Alignment and Applications to Wireless Interference Networks, *IEEE Transactions on Information Theory* Vol. 57(No. 6): 3309–3322.

[16] Hassibi, B. and Sharif, M. (2007). Fundamental limits in MIMO broadcast channels, *IEEE Journal on Selected Areas in Communications* Vol. 25(No. 7): 1333–1344.

[17] Duplicy, J., Badic, B. and Balraj, R. (2011). MU-MIMO in 4G systems, *Eurasip Journal on Wireless Communications and Networking* Vol. 2011: 1–12.

[18] Frank, P., Müller, A. and Speidel, J. (2010). Performance of CSI-based multi-user MIMO for the LTE downlink, *Proceedings of* 6th International Wireless Communications and Mobile Computing Conference (IWCMC), USA, pp. 1086–1090.

[19] Frank, P., Müller, A., Droste, H. and Speidel, J. (2010). Cooperative interference-aware joint scheduling for the 3GPP LTE uplink, *Proceedings of* IEEE 21st International Symposium on Personal Indoor and Mobile Radio Communications (PIMRC), pp. 2216–2221.

[20] Benjebbour, A., Shirakabe, M., Ohwatari, Y., Hagiwara, J. and Ohya, T. (2008). Evaluation of user throughput for MU-MIMO coordinated wireless networks, *Proceedings of* IEEE 19th International Symposium on Personal, Indoor and Mobile Radio Communications (PIMRC), pp. 1–5.

[21] Gesbert, D., Hanly, S., Huang, H., Shamai,S. and Yu, W.(2010). Multi-Cell MIMO Cooperative Networks: A New Look at Interference, *IEEE Journal on Selected Areas in Communications* Vol. 28(No. 9): 1380–1408.

[22] Spencer, Q. H. and Peel, C. B. and Swindlehurst, A. L. and Haardt, M. (2004). An introduction to the multi-user MIMO downlink, *IEEE Communications Magazine* Vol. 42(No. 10): 60–67.

[23] Weingarten, H., Steinberg, Y. and Shamai, S. (2006). The Capacity Region of the Gaussian MIMO Broadcast channel, *IEEE Journal on Selected Areas in Communications* Vol. 52(No. 9): 3936–3964.

[24] Hemrungrote, S., Hori, T., Fujimoto, M. and Nishimori, K. (2010). Channel Capacity Characteristics of Multi-User MIMO Systems in Urban Area, *Proceedings of* 2010 IEEE Antennas and Propagation Society International Symposium (APSURSI), pp. 1–4.

[25] Tse, D. and Viswanath, P. (2005). Fundamentals of wireless communication, Cambridge University Press, ISBN: 0–521–84527–0.

[26] Ravindran, N. and Jindal, N. (2007). Limited feedback-based block diagonalization for the MIMO broadcast channel, *IEEE Journal on Selected Areas in Communications*Vol. 26(No.8): 1473–1482.

[27] Zhang, J. and Andrews, J. G., and Heath, R. W. (2009) . Block Diagonalization in the MIMO Broadcast Channel with Delayed CSIT, *Proceedings of* IEEE Global Telecommunications Conference (GLOBECOM), Hawaii, pp. 1–6.

[28] Zhang ,R. (2010). Cooperative Multi-Cell Block Diagonalization with Per-Base-Station Power Constraints, *Proceedings of* IEEE Wireless Communications and Networking Conference (WCNC), pp. 1–6.

[29] Spencer, Q.H., Swindlehurst, A.L.and Haardt, M. (2004). Zero-forcing methods for downlink spatial multiplexing in multiuser MIMO channels, *IEEE Transactions on Signal Processing*Vol. 52(No.2): 461–471.

[30] Lee, J. and Jindal, N. (2007). High SNR Analysis for MIMO Broadcast Channels: Dirty Paper Coding versus Linear Precoding, *IEEE Transactions on Information Theory* Vol. 53(No.12): 4787–4792.

[31] Tarighat, A., Sadek, M. and Sayed, A.H. (2005). A multi user beamforming scheme for downlink MIMO channels based on maximizing signal-to-leakage ratios, *Proceedings of* IEEE International Conference on Acoustics, Speech, and Signal Processing (ICASSP), Philadelphia, pp. iii/1129–iii/1132.

[32] Mundarath, J.C. and Kotecha, J.H. (2008). Multi-User Multi-Input Multi-Output (MU-MIMO) Downlink Beamforming Systems with Limited Feedback, *Proceedings of* IEEE Global Telecommunications Conference (GLOBECOM), pp. 1–6.

Section 2

Random Beamforming in Multi-User MIMO Systems

Chapter 2

Random Beamforming in Multi – User MIMO Systems

Hieu Duy Nguyen, Rui Zhang and Hon Tat Hui

Additional information is available at the end of the chapter

http://dx.doi.org/10.5772/57131

1. Introduction

Wireless communication paradigm has evolved from single-user single-input single-output (SISO) and multiple-input multiple-output (MIMO) systems to multi-user (MU) MIMO counterparts, which are shown greatly improving the rate performance by transmitting to multiple users simultaneously. The sum-capacity and the capacity region of a single-cell MU MIMO downlink system or the so-called MIMO broadcast channel (MIMO-BC) can be attained by the nonlinear "Dirty Paper Coding (DPC)" scheme [1] [2] [3]. However, DPC requires a high implementation complexity due to the non-linear successive encoding/decoding at the transmitter/receiver, and is thus not suitable for real-time applications. Other studies have proposed to use alternative linear precoding schemes for the MIMO-BC, e.g., the block-diagonalization scheme [4], to reduce the complexity. More information on the key developments of single-cell MIMO communication can be found in, for example, [5] [6] [7].

The performance of a multi-cell MIMO-BC setup, however, is not well understood. It is worth noting that the multi-cell downlink system can be modelled in general as an interference channel (IC) setup. Characterization of the capacity region of the Gaussian IC is still an open problem even for the two-user case [8]. An important development recently is the so-called "interference alignment (IA)" transmission scheme (see, e.g., [9] and references therein). With the aid of IA, the maximum achievable degrees of freedom (DoF), in which the DoF is defined as the sum-rate normalized by the logarithm of the signal-to-noise ratio (SNR) as the SNR goes to infinity or the so-called "pre-log" factor, has been obtained for various ICs to provide useful insights on designing the optimal transmission for interference-limited MU systems. Besides IA-based studies for the high-SNR regime, there is a vast body of works in the literature which investigated the multi-cell cooperative downlink precoding/beamforming at a given finite user's SNR. These works are typically categorized based on two different assumptions on the cooperation level of base stations (BSs). For the case of "fully cooperative" multi-cell systems with global transmit message sharing across all the BSs, a virtual MIMO-BC channel is equivalently formed. Therefore, existing single-cell downlink precoding techniques can be applied (see, e.g., [10] [11] [12] and references therein) with a non-trivial modification to deal with the per-BS power constraints as compared to the conventional sum-power constraint for the single-cell MIMO-BC case. In contrast, if transmit messages are only locally known at each BS, coordinated precoding/beamforming can be implemented among BSs to control the inter-cell interference (ICI) to their best effort [13] [14]. In [15] [16] [17], various parametrical characterizations of the Pareto boundary of the achievable rate region have been obtained for the multiple-input single-output (MISO) IC with coordinated transmit beamforming and single-user detection.

More important, most of the proposed precoding schemes, in single- or multi-cell case, rely on the assumption of perfect channel state information (CSI) for all the intra- and inter-cell links at the transmitter, which may not be realistic in practical cellular systems with a large number of users. Consequently, the study of quantized channel feedback for the MIMO BC has been recently a very active area of research (see, e.g., [18] and references therein).

The single-beam "opportunistic beamforming (OBF)" and multi-beam "random beamforming (RBF)" schemes for the single-cell MISO-BC, introduced in [19] and [20], respectively, therefore attract a lot of attention since they require only *partial* CSI feedback to the BS. The fundamental idea is to exploit the multi-user diversity gain, by employing *opportunistic user scheduling*, to combat the inter-beam interferences. The achievable sum-rate with RBF in a single-cell system has been shown to scale identically to that with the optimal DPC scheme assuming perfect CSI as the number of users goes to infinity, for any given user's signal-to-noise ratio (SNR) [20] [21]. Essentially, the result implies that the intra-cell interference in a single-cell RBF system can be completely eliminated when the number of users are sufficiently large, and an "interference-free" MU broadcast system is attainable. This thus shows the optimality of the single-cell RBF and motives other studies on opportunistic communication.

Although substantial subsequent investigations and/or extensions of the single-cell RBF have been pursued, there are very few works on the performance of the RBF scheme in a more realistic multi-cell setup, where the ICI becomes a dominant factor. It is worth noting that as the universal frequency reuse becomes more favourable in future generation cellular systems, ICI becomes a more serious issue as compared to the traditional case with only a fractional frequency reuse.

One objective of this chapter is to present a literature survey on the vast body of works studying the single-cell OBF/RBF. The main purpose, however, is to introduce the recent investigations on multi-cell RBF systems. In this chapter, we first review the achievable rates of multi-cell RBF in finite-SNR regime. Such results, albeit important, do not provide any insight to the impact of the interferences on the system throughput. This motivates us to introduce the high-SNR/DoF analysis proposed in [22] and [23], which is useful in characterizing the performance of RBF under multi-user diversity and interference effects. Furthermore, it provides new insights on the role of *spatial receive diversity* in RBF, which is not well understood so far. It is revealed that receive diversity is significantly beneficial to the rate performance of multi-cell RBF systems [24]. This conclusion, interestingly, sharply contrasts with one based on the traditional asymptotic analysis, i.e., assuming that the number of users goes to infinity for any given user's SNR.

The remainder of this chapter is organized as follows. Section 2 describes the multi-cell MIMO-BC system model and the MISO/MIMO RBF schemes. Section 3 presents a literature survey on the single-cell OBF/RBF. Section 4 investigates the multi-cell RBF under finite-SNR regime. The high-SNR/DoF analysis is introduced in Section 5, based on which we study the interplay between the multi-user diversity, spatial receive diversity and spatial multiplexing gains achievable in RBF. Finally, Section 6 ends the chapter with some concluding remarks.

Notations: Scalars, vectors, and matrices are denoted by lower-case, bold-face lower-case, and bold-face higher-case letters, respectively. The transpose and conjugate transpose operators are denoted as $(\cdot)^T$ and $(\cdot)^H$, respectively. $\mathbb{E}[\cdot]$ denotes the statistical expectation. $\mathbf{Tr}(\cdot)$ represents the trace of a matrix. The distribution of a circularly symmetric complex Gaussian random variable with zero mean and covariance σ^2 is denoted by $\mathcal{CN}(0,\sigma^2)$; and $\sim$ stands for "distributed as". $\mathbb{C}^{x\times y}$ denotes the space of $x \times y$ complex matrices.

2. System model

This section introduces the multi-cell system model and RBF schemes used throughout this chapter. Particularly, we consider a C-cell MIMO-BC system with K_c mobile stations (MSs) in the c-th cell, $c = 1, \cdots, C$. For the ease of analysis, we assume that all BSs/MSs have the same number of transmit/receive antennas, denoted as N_T and N_R, respectively. We also assume a "homogeneous" channel setup, in which the signal and ICI powers between users of one cell are identical. At each communication time, the c-th BS transmits $M_c \leq N_T$ orthonormal beams with M_c antennas and selects M_c from K_c users for transmission. Suppose that the channels are flat-fading and constant over each transmission period of interest. The received signal of user k in the c-th cell is

$$\boldsymbol{y}_k^{(c)} = \boldsymbol{H}_k^{(c,c)} \sum_{m=1}^{M_c} \boldsymbol{\phi}_m^{(c)} s_m^{(c)} + \sum_{l=1,\, l\neq c}^{C} \sqrt{\gamma_{l,c}} \boldsymbol{H}_k^{(l,c)} \sum_{m=1}^{M_l} \boldsymbol{\phi}_m^{(l)} s_m^{(l)} + \boldsymbol{z}_k^{(c)}, \quad (1)$$

where $\boldsymbol{H}_k^{(l,c)} \in \mathbb{C}^{N_R\times M_l}$ denotes the channel matrix between the l-th BS and the k-th user of the c-th cell, which consists of independent and identically distributed (i.i.d.) $\sim \mathcal{CN}(0,1)$ elements; $\boldsymbol{\phi}_m^{(c)} \in \mathbb{C}^{M_c\times 1}$ and $s_m^{(c)}$ are the m-th randomly generated beamforming vector of unit norm and transmitted data symbol from the c-th BS, respectively; it is assumed that each BS has the total sum power, P_T, i.e., $\mathbf{Tr}\left(\mathbb{E}[\mathbf{s_c s_c^H}]\right) \leq P_T$, where $\mathbf{s}_c = [s_1^{(c)}, \cdots, s_{M_c}^{(c)}]^T$; $\gamma_{l,c} < 1$ stands for the signal attenuation from the l-th BS to any user of the c-th cell, $l \neq c$; and $\boldsymbol{z}_k^{(c)} \in \mathbb{C}^{N_R\times 1}$ is the additive white Gaussian noise (AWGN) vector, of which each element is $\sim \mathcal{CN}(0,\sigma^2)$, $\forall k, c$. In the c-th cell, the total SNR, the SNR per beam, and the interference-to-noise ratio (INR) per beam from the l-th cell, $l \neq c$, are denoted as $\rho = P_T/\sigma^2$, $\eta_c = P_T/(M_c\sigma^2)$, and $\mu_{l,c} = \gamma_{l,c}P_T/(M_l\sigma^2)$, respectively.

2.1. Multi-cell MISO RBF

In this case, $N_R = 1$ and the system model in (1) reduces to

$$y_k^{(c)} = \boldsymbol{h}_k^{(c,c)} \sum_{m=1}^{M_c} \boldsymbol{\phi}_m^{(c)} s_m^{(c)} + \sum_{l=1,\, l\neq c}^{C} \sqrt{\gamma_{l,c}} \boldsymbol{h}_k^{(l,c)} \sum_{m=1}^{M_l} \boldsymbol{\phi}_m^{(l)} s_m^{(l)} + z_k^{(c)}, \tag{2}$$

We consider a multi-cell RBF scheme, in which all BSs in different cells are assumed to be able to implement the conventional single-cell RBF [20] in a time synchronized manner, which is described as follows:

- In the training phase, the c-th BS generates M_c orthonormal beams, $\boldsymbol{\phi}_1^{(c)}, \cdots, \boldsymbol{\phi}_{M_c}^{(c)}$, and uses them to broadcast the training signals to all users in the c-th cell. The total power of each BS is assumed to be distributed equally over M_c beams.
- Feedback scheme: Each user in the c-th cell measures the signal-to-interference-plus-noise ratio (SINR) value for each of M_c beams (shown in (3) below), and feeds this scalar value together with the corresponding beam index back to the BS.

$$\begin{aligned}\mathrm{SINR}_{k,m}^{(c)} &= \frac{\frac{P_T}{M_c}\left|\boldsymbol{h}_k^{(c,c)}\boldsymbol{\phi}_m^{(c)}\right|^2}{\sigma^2 + \frac{P_T}{M_c}\sum_{i=1,i\neq m}^{M_c}\left|\boldsymbol{h}_k^{(c,c)}\boldsymbol{\phi}_i^{(c)}\right|^2 + \sum_{l=1,l\neq c}^{C}\gamma_{l,c}\frac{P_T}{M_l}\sum_{i=1}^{M_l}\left|\boldsymbol{h}_k^{(l,c)}\boldsymbol{\phi}_i^{(l)}\right|^2} \\ &= \frac{\eta_c\left|\boldsymbol{h}_k^{(c,c)}\boldsymbol{\phi}_m^{(c)}\right|^2}{1+\eta_c\sum_{i=1,i\neq m}^{M_c}\left|\boldsymbol{h}_k^{(c,c)}\boldsymbol{\phi}_i^{(c)}\right|^2 + \sum_{l=1,l\neq c}^{C}\mu_{l,c}\sum_{i=1}^{M_l}\left|\boldsymbol{h}_k^{(l,c)}\boldsymbol{\phi}_i^{(l)}\right|^2},\end{aligned} \tag{3}$$

where $m = 1, \cdots, M_c$.

- Opportunistic scheduling: The c-th BS schedules transmission to a set of M_c users for each time by assignning its m-th beam to the user with the highest SINR, i.e.,

$$k_m^{(c)} = \arg \max_{k\in\{1,\cdots,K_c\}} \mathrm{SINR}_{k,m}^{(c)}. \tag{4}$$

Then, the achievable average sum-rate in bits-per-second-per-Hz (bps/Hz) of the c-th cell is given by

$$R_{\mathrm{RBF}}^{(c)} = \mathbb{E}\left[\sum_{m=1}^{M_c} \log_2\left(1+\mathrm{SINR}_{k_m^{(c)},m}^{(c)}\right)\right] = M_c\mathbb{E}\left[\log_2\left(1+\mathrm{SINR}_{k_1^{(c)},1}^{(c)}\right)\right]. \tag{5}$$

Note that a different feedback scheme for single-cell RBF systems is considered in [20]. Sharif *et. al.* assumes that each user can only sends its maximum SINR, i.e., $\max_{1\leq m\leq M} \mathrm{SINR}_{k,m}$, along with the index m in which the SINR is maximized. The objective for introducing this scheme is to get a fair comparison with OBF. In this case, (5) is only a close approximation of the c-th sum rate by ignoring the small probability that one user may be assigned more than one beams for transmission and diminishes to zero as $K_c \to \infty$ [20]. In this chapter, the modified feedback scheme is considered since the focus is the rate performance under multi-user diversity, spatial receive diversity, and interference-limited effects.

2.2. Multi-cell MIMO RBF schemes

With multiple antennas, the MSs can apply receive diversity techniques to enhance the performance. In this chapter, we consider the three receiver designs, i.e., minimum-mean-square-error (MMSE), match-filter (MF), and antenna selection (AS), respectively. The necessity of employing such suboptimal schemes arises when MSs are constrained by the time-delay criterion or the complexity of the MMSE receiver. We formally define the multi-cell RBF with MMSE, MF, and AS receiver, denoted as RBF-MMSE, RBF-MF, and RBF-AS, respectively, as follows:

1. Training phase:

   a) The c-th BS generates M_c orthonormal beams, $\boldsymbol{\phi}_1^{(c)}, \cdots, \boldsymbol{\phi}_{M_c}^{(c)}$, and uses them to broadcast the training signals to all users in all cells. The total power of the c-th BS is assumed to be distributed equally over M_c beams.

   b1) MMSE: For the m-th beam, user k in the c-th cell does the following steps:

      i. Estimate the effective channel through training with the c-th BS: $\sqrt{\frac{P_T}{M_c}}\tilde{\boldsymbol{h}}_{k,m}^{(c,c)} = \sqrt{\frac{P_T}{M_c}}\boldsymbol{H}_k^{(c,c)}\boldsymbol{\phi}_m^{(c)}$.

      ii. Estimate the interference-plus-noise covariance matrix through training with all BSs:

$$\mathbf{W}_k^{(c)} = \frac{P_T}{M_c}\tilde{\boldsymbol{H}}_{k,-m}^{(c,c)}\left(\tilde{\boldsymbol{H}}_{k,-m}^{(c,c)}\right)^H + \sum_{l=1,l\neq c}^{C}\frac{P_T\gamma_{l,c}}{M_l}\tilde{\boldsymbol{H}}_k^{(l,c)}\left(\tilde{\boldsymbol{H}}_k^{(l,c)}\right)^H + \sigma^2\boldsymbol{I}, \tag{6}$$

      in which $\tilde{\boldsymbol{H}}_{k,-m}^{(c,c)} = \boldsymbol{H}_k^{(c,c)}[\boldsymbol{\phi}_1^{(c)}, \cdots, \boldsymbol{\phi}_{m-1}^{(c)}, \boldsymbol{\phi}_{m+1}^{(c)}, \cdots, \boldsymbol{\phi}_{M_c}^{(c)}]$, and $\tilde{\boldsymbol{H}}_k^{(l,c)} = \boldsymbol{H}_k^{(c,c)}[\boldsymbol{\phi}_1^{(l)}, \cdots, \boldsymbol{\phi}_{M_l}^{(l)}]$.

      iii. Apply the MMSE receive beamformer, i.e., $\boldsymbol{t}_{k,m}^{(c)} = \sqrt{\frac{P_T}{M_c}}\left(\mathbf{W}_k^{(c)}\right)^{-1}\tilde{\boldsymbol{h}}_{k,m}^{(c,c)}$, then calculate and feedback the SINR corresponding to the m-th transmit beam $\boldsymbol{\phi}_m^{(c)}$ to the BS

$$\text{SINR}_{k,m}^{(\text{MMSE},c)} = \frac{P_T}{M_c}\left(\tilde{\boldsymbol{h}}_{k,m}^{(c,c)}\right)^H\left(\mathbf{W}_k^{(c)}\right)^{-1}\tilde{\boldsymbol{h}}_{k,m}^{(c,c)}. \tag{7}$$

   b2) MF: For the m-th beam, user k in the c-th cell does the following steps:

      i. Estimate the effective channel through training with the c-th BS: $\sqrt{\frac{P_T}{M_c}}\tilde{\boldsymbol{h}}_{k,m}^{(c,c)} = \sqrt{\frac{P_T}{M_c}}\boldsymbol{H}_k^{(c,c)}\boldsymbol{\phi}_m^{(c)}$.

      ii. Apply the MF, i.e., $\boldsymbol{t}_{k,m}^{(c)} = \tilde{\boldsymbol{h}}_{k,m}^{(c,c)} \;/\; ||\tilde{\boldsymbol{h}}_{k,m}^{(c,c)}||$. The rationale is to maximize the power received from the desired beam. The receive signal now is

$$r_{k,m}^{(c)} = \left(\boldsymbol{t}_{k,m}^{(c)}\right)^H \boldsymbol{y}_k^{(c)} = \sqrt{\frac{P_T}{M_c}} \left(\boldsymbol{t}_{k,m}^{(c)}\right)^H \tilde{\boldsymbol{h}}_{k,m}^{(c,c)} s_m^{(c)} + \sqrt{\frac{P_T}{M_c}} \left(\boldsymbol{t}_{k,m}^{(c)}\right)^H \tilde{\boldsymbol{H}}_{k,-m}^{(c,c)} \boldsymbol{s}_{-m}^{(c)}$$
$$+ \sum_{l=1,l\neq c} \sqrt{\frac{P_T \gamma_{l,c}}{M_l}} \left(\boldsymbol{t}_{k,m}^{(c)}\right)^H \tilde{\boldsymbol{H}}_k^{(l,c)} \boldsymbol{s}_{(l)} + \left(\boldsymbol{t}_{k,m}^{(c)}\right)^H \boldsymbol{z}_k^{(c)}, \quad (8)$$

where $s_{-m}^{(c)} = [\boldsymbol{s}_1^{(c)}, \cdots, \boldsymbol{s}_{m-1}^{(c)}, \boldsymbol{s}_{m+1}^{(c)}, \cdots, \boldsymbol{s}_{M_c}^{(c)}]$ and $s_m^{(l)} = [\boldsymbol{s}_1^{(l)}, \cdots, \boldsymbol{s}_{M_l}^{(c)}]$. Through training with all BSs, user k in the c-th cell estimates the interferences' power in (8), which can be expressed equivalently as $\left(\tilde{\boldsymbol{h}}_{k,m}^{(c,c)}\right)^H \left(\boldsymbol{W}_k^{(c)}\right) \tilde{\boldsymbol{h}}_{k,m}^{(c,c)}$, in which $\boldsymbol{W}_k^{(c)}$ is defined as in (6). Each user then calculates and feedbacks the SINR corresponding to the m-th transmit beam $\boldsymbol{\phi}_m^{(c)}$ to the BS, which can be expressed equivalently as follows

$$\text{SINR}_{k,m}^{(\text{MF},c)} = \frac{\frac{P_T}{M_c} ||\tilde{\boldsymbol{h}}_{k,m}^{(c,c)}||^4}{\left(\tilde{\boldsymbol{h}}_{k,m}^{(c,c)}\right)^H \left(\boldsymbol{W}_k^{(c)}\right) \tilde{\boldsymbol{h}}_{k,m}^{(c,c)}}, \quad (9)$$

Note that the k-th user in the c-th cell knows $\left(\tilde{\boldsymbol{h}}_{k,m}^{(c,c)}\right)^H \left(\boldsymbol{W}_k^{(c)}\right) \tilde{\boldsymbol{h}}_{k,m}^{(c,c)}$ but not $\boldsymbol{W}_k^{(c)}$.

b3) AS: The received signal at the n-th receive antenna of user k in the c-th cell is:

$$y_{k,n}^{(c)} = \boldsymbol{h}_{k,n}^{(c,c)} \sum_{m=1}^{M_c} \boldsymbol{\phi}_m^{(c)} s_m^{(c)} + \sum_{l=1,\, l\neq c}^{C} \sqrt{\gamma_{l,c}} \boldsymbol{h}_{k,n}^{(l,c)} \sum_{m=1}^{M_l} \boldsymbol{\phi}_m^{(l)} s_m^{(l)} + z_{k,n}^{(c)}, \quad (10)$$

where $y_{k,n}^{(c)}$ and $z_{k,n}^{(c)}$ are the n-th element of $\boldsymbol{y}_k^{(c)}$ and $\boldsymbol{z}_k^{(c)}$, respectively, and $\boldsymbol{h}_{k,n}^{(l,c)} \in \mathbb{C}^{1\times M_l}$ is the n-th row of $\boldsymbol{H}_k^{(l,c)}$, $n \in \{1, \ldots, N_R\}$, $l, c \in \{1, \ldots, C\}$. For the m-th beam and n-th antenna, user k estimates the received power of the signal and the interferences through training with the c-th BS and all BSs, respectively. The following SINR value is then calculated:

$$\text{SINR}_{k,n,m}^{(\text{AS},c)} = \frac{\frac{P_T}{M_c} \left|\boldsymbol{h}_{k,n}^{(c,c)} \boldsymbol{\phi}_m^{(c)}\right|^2}{\sigma^2 + \frac{P_T}{M_c} \sum_{i=1,i\neq m}^{M_c} \left|\boldsymbol{h}_{k,n}^{(c,c)} \boldsymbol{\phi}_i^{(c)}\right|^2 + \sum_{l=1,l\neq c}^{C} \gamma_{l,c} \frac{P_T}{M_l} \sum_{i=1}^{M_l} \left|\boldsymbol{h}_{k,n}^{(l,c)} \boldsymbol{\phi}_i^{(l)}\right|^2}. \quad (11)$$

Denote

$$\text{SINR}_{k,m}^{(\text{AS},c)} := \max_{n\in\{1,\cdots,N_R\}} \text{SINR}_{k,n,m}^{(\text{AS},c)}. \quad (12)$$

For each m-th beam, the k-th user selects antenna $n_{k,m}^{(c)}$, which corresponds to $\mathrm{SINR}_{k,m}^{(\mathrm{AS},c)}$, and feeds $\mathrm{SINR}_{k,m}^{(\mathrm{AS},c)}$ back to the BS.

2. Scheduling phase: The c-th BS assigns the m-th beam to the user with the highest SINR

$$k_m^{(\mathrm{Rx},c)} = \arg \max_{k \in \{1,\cdots,K_c\}} \mathrm{SINR}_{k,m}^{(\mathrm{Rx},c)}, \tag{13}$$

in which "Rx" denotes either MMSE, MF, or AS.

3. Transmitting phase: The c-th BS uses the m-th beam to transmit data to user $k_m^{(\mathrm{Rx},c)}$.

The achievable sum-rate in bits per second per Hz (bps/Hz) of the c-th RBF-Rx cell can be expressed as

$$R_{\text{RBF-Rx}}^{(c)} = \mathbb{E}\left[\sum_{m=1}^{M_c} \log_2\left(1 + \mathrm{SINR}_{k_m^{(\mathrm{Rx},c)},m}^{(\mathrm{Rx},c)}\right)\right] = M_c \mathbb{E}\left[\log_2\left(1 + \mathrm{SINR}_{k_1^{(\mathrm{Rx},c)},1}^{(\mathrm{Rx},c)}\right)\right]. \tag{14}$$

Remark 2.1. *The RBF-AS scheme consists of two selection processes: antenna selection at the MSs each with N_R antennas and opportunistic scheduling at the BSs with K_c's users. The rate performance of RBF-AS is therefore equal to that of the (MISO-BC) RBF with $N_R K_c$ single-antenna users in the c-th cell. The RBF-AS scheme is introduced here to provide a complete study and rigorous comparison.*

3. Literature survey on single-cell OBF/RBF

Since its introduction in the landmark paper [19], opportunistic communication has developed to a broad area with various constituent topics. In this section, we aim to present a succinct overview on the key developments of OBF/RBF, summarizing some of the most important results contributed to the field. Note that in the literature, virtually all the works consider the single-cell case. It is only quite recent that the rate performance of the multi-cell RBF and ad-hoc IC with user scheduling is explored in [24] [22] [23]. We therefore limit our survey to the single-cell OBF/RBF. In this section, the channel model (2) has $C = 1$ and the cell index c is dropped for brevity. (3) and (5) hence reduce to

$$\mathrm{SINR}_{k,m} = \frac{\frac{P_T}{M}\left|\boldsymbol{h}_k \boldsymbol{\phi}_m\right|^2}{\sigma^2 + \frac{P_T}{M}\sum_{i=1,i\neq m}^{M}\left|\boldsymbol{h}_k \boldsymbol{\phi}_i\right|^2}. \tag{15}$$

$$R_{\mathrm{RBF}} = M\mathbb{E}\left\{\log_2\left(1 + \max_{k\in\{1,\cdots,K\}} \mathrm{SINR}_{k,1}\right)\right]. \tag{16}$$

The probability density function (PDF) and cumulative density function (CDF) of $S := \mathrm{SINR}_{k,m}, \forall k, m$ can be expressed as [20]

$$f_S(s) = \frac{e^{-s/\eta}}{(s+1)^M}\left(M - 1 + \frac{s+1}{\eta}\right). \tag{17}$$

$$F_S(s) = 1 - \frac{e^{-s/\eta}}{(s+1)^{M-1}}. \tag{18}$$

where $\eta = P_T/(M\sigma^2)$ is the SNR per beam.

3.1. Achievable rate

With the feedback scheme stated in Section 2.1, the closed-form expression for the sum rate R_{RBF} is given in the following lemma.

Lemma 3.1. *([22, Lemma 3.1], see also [55]) The average sum rate of the single-cell RBF is given by*

$$R_{RBF} = \frac{M}{\log 2} \sum_{n=1}^{K} (-1)^n \binom{K}{n} \left[\left(-\frac{n}{\eta}\right)^{n(M-1)} \frac{e^{n/\eta} Ei(-n/\eta)}{(n(M-1))!} - \sum_{m=1}^{n(M-1)} \left(-\frac{n}{\eta}\right)^{m-1} \frac{(n(M-1)-m)!}{(n(M-1))!} \right], \tag{19}$$

where $Ei(x) = \int_{-\infty}^{x} \frac{e^t}{t} dt$ *is the exponential integral function.*

Assuming the feedback scheme in [20], (19) becomes a close approximation of the exact sum rate, especially when the number of users K is large. The exact expression of the sum rate is derived in [25], which involves a numerical integral with the SINR CDF of the first beam:

Lemma 3.2. *([25]) Assuming the feedback scheme in [20], the average sum rate of the single-cell RBF is given by*

$$R'_{RBF} = M \int_0^\infty \log_2(1+x) f'_{SINR_{k_1,1}}(x) dx, \tag{20}$$

in which

$$f'_{SINR_{k_1,1}}(x) = \sum_{k=1}^{K} \frac{1}{M} \left(1 - \frac{1}{M}\right)^{k-1} \frac{K!}{(K-k)!(k-1)!} (F_1(x))^{K-k} [1 - F_1(x)]^{k-1} f_1(x) + \left(1 - \frac{1}{M}\right)^{K} \frac{1}{M-1} \sum_{m=2}^{M} f_{2,m}(x), \tag{21}$$

$$F_1(x) = \int_0^x f_1(t) dt, \tag{22}$$

$$f_1(x) = \sum_{m=0} M - 1 \frac{(-1)^m M!}{(M-m-1)!m!} \exp\left(-\frac{(1+m)x}{\eta(mx-1)}\right) \times \frac{(M + 1/\eta - 1 + (1/\eta - mM + m)x)(1-mx)^{M-3}}{(1+x)^M}, \tag{23}$$

http://dx.doi.org/10.5772/57131

$$f_{2,m}(x) = \int_0^{\infty} \int_0^{\frac{(M-m)w}{m-1}} \left(\frac{1}{\eta} + z + w\right) f_{w_m,\beta_{m:M},z_m}\left(w, x\left(\frac{1}{\rho} + z + w\right), z\right) dzdw. \quad (24)$$

Finally, the function $f_{w_m,\beta_{m:M},z_m}(w, \beta, z)$ *is given in [26, (28)]*

$$\begin{aligned} f_{w_m,\beta_{m:M},z_m}(w,\beta,z) = & \frac{M![w-(m-1)\beta]^{m-1}}{(L-1)!(l-1)!\bar{\beta}^M} \frac{\exp\left(-\frac{w+\beta+z}{\bar{\beta}}\right)}{(m-2)!(M-m-1)!} \mathcal{U}(1-(m-1)\beta) \times \\ & \times \sum_{i=0}^{M-m} \binom{M-m}{i} (-1)^i (z-i\beta)^{M-m-1} \mathcal{U}(z-i\beta), \\ \beta > 0, w > (m-1)\beta, z < (M-1)\beta, & \quad (25) \end{aligned}$$

where $\mathcal{U}(x)$ *is the unit step function and* $\bar{\beta} = 1$ *due to the Rayleigh fading channel model.*

In addition, loose approximations for (5) are presented in [27] and [28]. We note that (19) involves only the exponential integral function, which is more efficiently computable than the Gaussian hyper-geometric functions in [27] [28] and the (exact) expression in Lemma 3.2. However, the sum-rate approximations in [27] and [28] can directly lead to some asymptotic results, e.g., the sum-rate scaling law $M \log_2 \log K$ as $K \to \infty$, while (19) and (20) do not.

3.2. Asymptotic analysis

The accurate expression for the achievable sum rate, albeit important, is too complicated. To reveal more insights on the performance of single-cell RBF, asymptotic analyses have been considered in other studies. There are two main approaches, namely, large-number-of-users and large-system analyses.

3.2.1. Large number of users

The conventional asymptotic investigation of OBF/RBF is to consider the number of users approaches infinity for a given finite SNR. Based on *extreme value theory* [29], one of the most important results in opportunistic communication is proved in [20][1].

Theorem 3.1. *([20, Theorem 1]) For fixed* $M \leq N_T$ *and* P_T*, the single-cell RBF sum rate grows double-logarithmically with respect to the number of users, i.e.,*

$$\lim_{K \to \infty} \frac{R_{RBF}}{M \log_2 \log K} = 1. \quad (26)$$

In [21], Sharif *et. al.* show that $N_T \log_2 \log K$ is also the rate scaling law of the optimal DPC scheme assuming perfect CSI as the number of users goes to infinity, for any given

[1] Strictly speaking, Theorem 3.1 only states for the single-cell RBF with the original feedback scheme in [20]. However, it is easy to see that the same result applies when the modified feedback scheme in Section 2.1 is considered.

user's SNR. Essentially, the result implies that the intra-cell interference in a single-cell RBF system can be completely eliminated when the number of users is sufficiently large, and an "interference-free" MU broadcast system is attainable. This important result therefore establishes the optimality of single-cell RBF and motivates further studies on opportunistic communication. Various MIMO-BC transmission schemes with different assumptions on the fading model, feedback scheme, user scheduling, etc., can be shown achieving $N_T \log_2 \log K$ as $K \to \infty$, suggesting that it is a very universal rate scaling law (see, e.g., [30] [31] [28] [32] [33] [34]). Finally, it is worth noting that the RBF sum rate grows only *logarithmically* with K, i.e., $R_{RBF} \,/\, \frac{M}{M-1} \log_2 K \to 1$ as $K \to \infty$, when the background noise is ignored [30] [33]. More discussions on the large-number-of-users analysis will be given in Section 4.1.2 and 4.2.

3.2.2. *Large system*

The large-system analysis is a well-known and widely-accepted method to investigate the performance of communication systems. A recent application to single-cell RBF is introduced in [35], in which a general MIMO-BC setup with MMSE receiver and different fading models is considered. Assuming the numbers of transmit/receive antennas and data beams to approach infinity at the same time with fixed ratios for any given finite SNR, Couillet *et. al.* obtain "almost closed-form" numerical solutions which provide deterministic approximations for various performance criteria. Note that although these results are derived under the large-system assumption, Monte-Carlo simulations demonstrate that they can be applied to study small-dimensional systems with modest errors. However, [35] does not consider opportunistic scheduling, which is one of the main features of RBF.

3.3. Reduced and quantized feedback in OBF/RBF

In practical systems, only a limited number of bits representing the quantized channel gain/SINR can be sent from each user to the corresponding BS. Note that the feedback schemes in Section 2.1 and [20] require the transmission of $2MK$ and $2K$ scalar values from K users, respectively, i.e., a linear increase with respect to the number of users. It is thus of great interest to develop schemes which can reduce the numbers of users and/or bits to be fed back. The idea of using only one-bit feedback is introduced in several works, e.g., [36] [37] [38]. In this scheme, the user sends "1" when the SINR value is above a pre-determined threshold, and "0" vice-versa. Since the performance of OBF/RBF only depends on the favourable channels, one bit of feedback per user can capture almost all gain available due to the multi-user diversity. Optimal quantization strategy for OBF systems with more than one bit feedback is proposed in [36]. It is also worth noting the group random access-based feedback scheme in [39] and the multi-user diversity/throughput tradeoff analysis in [40]. The main tool to study the performance of OBF/RBF under reduced feedback schemes is the large-number-of-users analysis.

3.4. Non-orthogonal RBF and Grassmanian line packing problem

Denote the space of unit-norm transmit beamforming vectors in $\mathbb{C}^{N_T \times 1}$ as $\mathbb{O}(N_T, 1)$. A distance function of v_1, $v_2 \in \mathbb{O}(N_T, 1)$ can be defined to be the sine of the angle between them [41]

$$d(v_1, v_2) = sin(v_1, v_2) = \sqrt{1 - |v_1^H v_2|^2}, \quad (27)$$

which is known as the chordal distance. The problem of finding the packing of M unit-norm vectors in $\mathbb{C}^{N_T \times 1}$ that has the maximum minimum distance between any pair of them is called the *Grassmannian line packing problem* (GLPP). The GLPP appears in the problem of designing beamforming codebook for space-time codes [42] [43].

Given that the number of transmit beams is less than or equal to the number of transmit antennas, i.e., $M \leq N_T$, any orthonormal set $\{\boldsymbol{\phi}_m\}_{m=1}^M$ is a solution of the GLPP. However, assuming that the BS sends $M > N_T$ beams to serve more users simultaneously and improve the fairness of the system, finding $\{\boldsymbol{\phi}_m\}_{m=1}^M$ is a nontrivial GLPP. The idea of using more than N_T beams is first proposed in [44] with $M = N_T + 1$, and further studied in [45]. Zorba *et. al.* argue that the scaling law $M \log_2 \log K$ is still true for $M = N_T + 1$ case [44], while the results in [45] imply that non-orthogonal beamforming matrix induces an interference-limited effect on the sum rate, and the multi-user diversity vanishes. Since both studies are based on approximated derivations, more rigorous investigations are necessary before any conclusion is drawn.

3.5. Other studies

Beam selection and beam power control algorithms for single-cell RBF are proposed in [46] [47] [33] [48]. The objective is to improve the rate performance especially when the number of users is not so large. The idea of employing a codebook of predetermined orthonormal beamforming matrices is introduced in [30] [31] [27]. While [30] [31] investigate RBF when quantized, normalized channel vectors are fed back to the BS, [49] studies the codebook design problem and the rate performance assuming that opportunistic selection is also performed on the codebook. These problems are related to Section 3.3 and the GLPP in Section 3.4. Fairness scheduling problem is studied in [19], in which the "*proportional fair scheduling* (PFS)" scheme is proposed. It is not surprising that most of the later developments approach PFS from a network layer's perspective. Notably, the convergence of PFS algorithm for many-user cases under general network conditions is proved in [50], and a global PFS for multi-cell systems is introduced in [51].

4. Multi-cell RBF: Finite-SNR analysis

4.1. Multi-cell MISO RBF

In this subsection, we review the recent results on the achievable sum rate of the multi-cell MISO RBF scheme under finite-SNR regime [22]. In particular, we first consider an extension of Lemma 3.1 to the multi-cell case subject to the ICI, and furthermore show the asymptotic sum-rate scaling law as the number of users per cell tends to infinity.

For the single-cell RBF case, the SINR distributions given in (17) and (18) are obtained in [20]. The following lemma establishes similar results for the multi-cell case

Lemma 4.1. *([22, Lemma 3.2], see also Section 4.2) In the multi-cell MISO RBF, the PDF and CDF of the SINR $S := \mathrm{SINR}_{k,m}^{(c)}$, $\forall k, m$, have closed-form expressions given by*

$$f_S^{(c)}(s) = \frac{e^{-s/\eta_c}}{(s+1)^{M_c-1} \prod_{l=1, l\neq c}^{C} \left(\frac{\mu_{l,c}}{\eta_c} s + 1\right)^{M_l}} \left[\frac{1}{\eta_c} + \frac{M_c - 1}{s+1} + \sum_{l=1, l\neq c}^{C} \frac{M_l}{s + \frac{\eta_c}{\mu_{l,c}}}\right], \tag{28}$$

$$F_S^{(c)}(s) = 1 - \frac{e^{-s/\eta_c}}{(s+1)^{M_c-1} \prod_{l=1, l\neq c}^{C} \left(\frac{\mu_{l,c}}{\eta_c} s + 1\right)^{M_l}}. \tag{29}$$

4.1.1. Sum rate with finite K

With Lemma 4.1, Lemma 3.1 is readily generalized to the multi-cell case in the following theorem.

Theorem 4.1. *([22, Theorem 3.1]) The total sum rate of a C-cell MISO RBF system equals to $\sum_{c=1}^{C} R_{RBF}^{(c)}$, in which the individual sum rate of the c-th cell, $R_{RBF}^{(c)}$, is given by*

$$\begin{aligned} R_{RBF}^{(c)} = {} & \frac{M_c}{\log 2} \sum_{n=1}^{K_c} (-1)^n \binom{K_c}{n} \prod_{l=1, l\neq c}^{C} \left(\frac{\eta_c}{\mu_{l,c}}\right)^{nM_l} \times \\ & \left\{ \sum_{p=1}^{n(M_c-1)+1} \frac{A_{n,c,p}}{(p-1)!} \left[e^{\frac{n}{\eta_c}} \left(-\frac{n}{\eta_c}\right)^{p-1} Ei\left(-\frac{n}{\eta_c}\right) - \sum_{m=1}^{p-1} \left(-\frac{n}{\eta_c}\right)^{m-1} (p-1-m)! \right] \right. \\ & + \sum_{l=1, l\neq c}^{C} \sum_{q=1}^{nM_l} \frac{A_{n,l,q}}{(q-1)!} \left[e^{\frac{n}{\mu_{l,c}}} \left(-\frac{n}{\eta_c}\right)^{q-1} Ei\left(-\frac{n}{\mu_{l,c}}\right) \right. \\ & \left.\left. - \sum_{m=1}^{q-1} \left(-\frac{n}{\eta_c}\right)^{m-1} \left(\frac{\mu_{l,c}}{\eta_c}\right)^{q-m} (q-1-m)! \right] \right\}, \end{aligned} \tag{30}$$

where $A_{n,c,p}$'s and $A_{n,l,q}$'s are the coefficients from the following partial fractional decomposition:

$$\frac{1}{(x+1)^{n(M_c-1)+1} \prod_{l\neq c}^{C} \left(x + \frac{\eta_c}{\mu_{l,c}}\right)^{nM_l}} = \sum_{p=1}^{n(M_c-1)+1} \frac{A_{n,c,p}}{(x+1)^p} + \sum_{l=1, l\neq c}^{C} \sum_{q=1}^{nM_l} \frac{A_{n,l,q}}{\left(x + \frac{\eta_c}{\mu_{l,c}}\right)^q}, \tag{31}$$

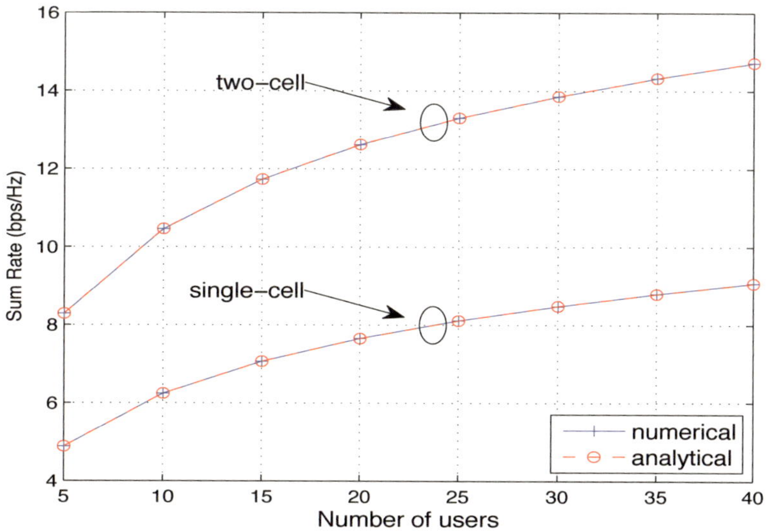

Figure 1. Comparison of the analytical and numerical results on the RBF sum rate.

and thus given by [52, (2.102)]:

$$A_{n,c,p} = \frac{1}{(n(M_c-1)-p+1)!} \frac{d^{n(M_c-1)-p+1}}{dx^{n(M_c-1)-p+1}} \left[\frac{1}{\prod_{l \neq c}^{C} \left(x + \frac{\eta}{\mu_l} \right)^{nM_l}} \right] \Bigg|_{x=-1}, \tag{32}$$

$$A_{n,l,q} = \frac{1}{(nM_l-q)!} \frac{d^{nM_l-q}}{dx^{nM_l-q}} \left[\frac{1}{(x+1)^{n(M_c-1)+1} \prod_{t \neq l,c}^{C} \left(x + \frac{\eta}{\mu_t} \right)^{nM_t}} \right] \Bigg|_{x=-\eta_c/\mu_{l,c}}. \tag{33}$$

If the original feedback scheme in [20] is employed, (30) becomes a close approximation of the c-th sum rate. Note that no other results than Theorem 4.1 are available in the literature for the multi-cell RBF sum rate.

Fig. 1 shows the analytical and numerical results on the RBF sum rate as a function of the number of users for both single-cell and two-cell systems. In the single-cell case, $M = N_T = 3$, $\eta = 20$ dB, while in the two-cell case, $K_1 = K_2$, $M_1 = M_2 = N_T = 3$, $\eta_1 = \eta_2 = 20$ dB, $\mu_{2,1} = 6$ dB, and $\mu_{1,2} = 10$ dB. It is observed that the sum-rate expressions in (19) and (30) are very accurate. Thus, it is possible to use Theorem 4.1 to compute all the sum-rate tradeoffs among different cells in a multi-cell RBF system, which leads to the achievable rate region. However, such a characterization requires intensive computations, and does not provide any useful insights.

4.1.2. Asymptotic sum rate as $K_c \to \infty$

An attempt to extend the famous scaling law $M \log_2 \log K$ to the multi-cell RBF case has been made in [53] based on an approximation of the SINR's PDF (which is applicable if

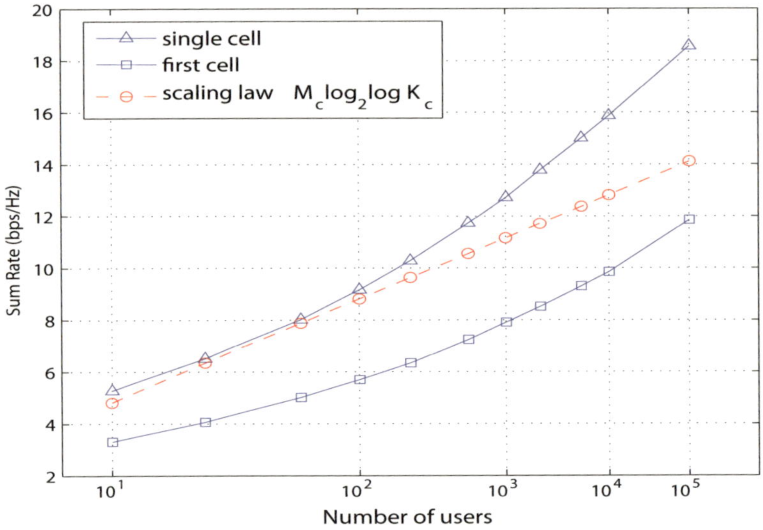

Figure 2. Comparison of the numerical sum rate and the sum-rate scaling law for RBF.

the SNR and INRs are all roughly equal), by showing that the same asymptotic sum-rate $M_c \log_2 \log K_c$ for each individual cell holds as the single-cell case. However, we note that with the exact SINR distributions in Lemma 4.1, a rigorous proof can be obtained, leading to the following proposition.

Proposition 4.1. *([22, Proposition 3.1]) For fixed M_c's and P_T, $c = 1, \cdots, C$, we have*

$$\lim_{K_c \to \infty} \frac{R_{RBF}^{(c)}}{M_c \log_2 \log K_c} = 1. \tag{34}$$

Fig. 2 depicts both the numerical and theoretical asymptotic sum-rates for a single-cell RBF system, and the first cell of a two-cell RBF system. In the single-cell case, $\eta = 10$ dB, and $M = N_T = 4$, while in the two-cell case, $M_1 = M_2 = N_T = 4$, $\eta_1 = 10$ dB, and $\mu_{2,1} = 6$ dB. We observe that the convergence to the sum-rate scaling law $M_c \log_2 \log K_c$ is rather slow in both cases. For example, even with K or K_c to be 10^5, the convergence is still not clear. In fact, Fig. 2 indicates that the sum-rate might follow a (single-) logarithmically scaling law. Furthermore, Proposition 4.1 implies that the sum-rate scaling law $M_c \log_2 \log K_c$ holds for any cell regardless of the ICI as $K_c \to \infty$. As a consequence, this result implies that each cell should apply the maximum number of transmit beams, i.e., $M_c = N_T, \forall c$, to maximize the per-cell throughput. Such a conclusion may be misleading in a practical multi-cell system with non-negligible ICI. The above two main drawbacks, namely, slow convergence and misleading conclusion, limit the usefulness of the conventional sum-rate scaling law $M_c \log_2 \log K_c$ for the multi-cell RBF.

4.2. Multi-cell MIMO RBF

To study the achievable sum-rates of the RBF-Rx schemes, it is necessary to investigate the SINRs given in (7), (9), and (12). The following lemmas establish the SINR distributions in three cases.

Lemma 4.2. *([24, Corollary 3.1], see also [54]) Given $N_R \leq \sum_{l=1}^{C} M_l - 1$, the CDF of the random variable $S := SINR_{k,m}^{(MMSE,c)}$ in (7) is*

$$F_S(s) = 1 - \frac{e^{-s/\eta_c}\left(\sum_{i=0}^{N_R-1} \zeta_i s^i\right)}{(1+s)^{M_c-1}\prod_{l=1,l\neq c}^{\sum M_c-1}(1+\frac{\mu_{l,c}}{\eta_c}s)^{M_l}}, \tag{35}$$

in which ζ_i is the coefficient of s^i in the product $(1+s)^{M_c-1}\prod_{l=1,l\neq c}^{\sum M_c-1}(1+\frac{\mu_{l,c}}{\eta_c}s)^{M_l}$.

Lemma 4.3. *([24, Theorem 3.2]) The CDF of $S := SINR_{k,m}^{(MF,c)}$ can be expressed as*

$$F_S(s) = 1 - e^{-s/\eta_c}\sum_{k=0}^{N_R-1}\sum_{j=0}^{k}\frac{(-1)^j s^k}{(k-j)!j!\eta_c^{k-j}}\frac{d^j T_0}{ds^j}, \tag{36}$$

in which

$$T_0 = \frac{1}{(1+s)^{M_c-1}\prod_{l=1,l\neq c}^{C}(1+\frac{\mu_{l,c}}{\eta_c}s)^{M_l}}. \tag{37}$$

Lemma 4.4. *([24, Corollary 3.2]) The CDF of $S := SINR_{k,m}^{(AS,c)}$ can be expressed as*

$$F_S(s) = \left(1 - \frac{e^{-s/\eta_c}}{(s+1)^{M_c-1}\prod_{l=1,l\neq c}^{C}\left(\frac{\mu_{l,c}}{\eta_c}s+1\right)^{M_l}}\right)^{N_R}. \tag{38}$$

In Fig. 3, we present the per-cell SINR CDFs of the RBF-MMSE, RBF-MF, and RBF-AS with the following setup: $C = 4$, $\eta_1 = 20$ dB, $N_R = 3$, $M_1 = 3$, $[\mu_{2,1}, \mu_{3,1}, \mu_{4,1}] = [0, -3, 3]$ dB, and $[M_2, M_3, M_4] = [3, 2, 4]$. The numerical results are obtained with Monte-Carlo simulations, while the analytical results are computed based on Lemma 4.2, 4.3, and 4.4. As a reference, we also present the numerical CDF of the MISO RBF with single-antenna users. It is observed that spatial receive diversity does help improving the SINR performance. However, only in the RBF-MMSE scheme that there exists a tremendous gain. Finally, we note that the SINR performance of the RBF-MF is *not* always better than that of the RBF-AS.

With Lemma 4.2, 4.3, and 4.4, it is possible to extend Theorem 4.1 to multi-cell MIMO RBF systems. The results posses complicated expressions and, again, does not lead to useful insights. We are more interested in the following proposition

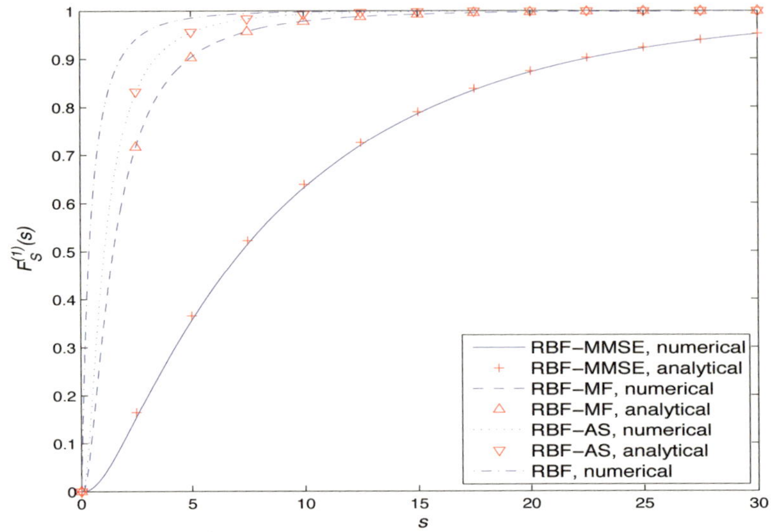

Figure 3. Comparison of the numerical and analytical CDFs of the per-cell SINR.

Proposition 4.2. *([24, Proposition 4.2]) For fixed M_c, N_R and P_T, $c = 1, \cdots, C$, we have*

$$\lim_{K_c \to \infty} \frac{R^{(c)}_{RBF\text{-}Rx}}{M_c \log_2 \log K_c} = 1, \tag{39}$$

where "Rx" denotes any receive diversity scheme at the users, e.g., MMSE, MF, or AS.

Essentially, this proposition extends Proposition 4.1 to the multi-cell MIMO case. Note that a similar result for a single-cell setup is obtained in [20] [21]. The proposition implies that spatial receive diversity only provides *marginal* gain to the rate performance of multi-cell MIMO RBF systems. However, as shown in Section 4.1.2, the traditional asymptotic analysis might not accurately characterize the RBF sum-rate in the condition of a small number of users (even up to K, $K_c = 10^5$). This negative conclusion on the receive diversity, therefore, might be misleading. The benefit of receive diversity to multi-cell RBF systems will be discussed further in Section 5.2.

5. Multi-cell RBF: High-SNR analysis and the degrees of freedom region

Motivated by the major limitations of the finite-SNR analysis, [22] and [23] propose using the high-SNR/DoF analysis to study the rate performance of multi-cell/IC systems with opportunistic scheduling. In particular, the multi-cell MISO RBF and ad-hoc SISO IC are considered in [22] and [23], respectively. A further extension is [24], in which the multi-cell MIMO RBF schemes are investigated to reveal the benefit of receive diversity. It has been shown that the high-SNR/DoF analysis provides a succinct description of the interplay between the multi-user diversity, spatial receive diversity and spatial multiplexing gains achievable in RBF.

In this section, we introduce the high-SNR/DoF analysis and the studies on multi-cell MISO/MIMO RBF. We first specify the DoF region, which was defined in [9] and is restated below for convenience.

Definition 5.1. *(General DoF region) The DoF region of a C-cell MIMO-BC system is defined as*

$$\mathcal{D}_{MIMO} = \Big\{ (d_1, d_2, \cdots, d_C) \in \mathbb{R}_+^C : \forall(\omega_1, \omega_2, \cdots, \omega_C) \in \mathbb{R}_+^C; \tag{40}$$
$$\sum_{c=1}^{C} \omega_c d_c \leq \lim_{\rho \to \infty} \sup_{\mathbf{R} \in \mathcal{R}} \sum_{c=1}^{C} \omega_c \frac{R_{sum}^{(c)}}{\log_2 \rho} \Big\},$$

where ρ is the per-cell SNR; ω_c, d_c, and $R_{sum}^{(c)}$ are the non-negative rate weight, the achievable DoF, and the sum rate of the c-th cell, respectively; and the region $\mathcal{R}$ is the set of all the achievable sum-rate tuples for all the cells, denoted by $\mathbf{R} = (R_{sum}^{(1)}, R_{sum}^{(2)}, \cdots, R_{sum}^{(C)})$.

If the multi-cell MIMO RBF schemes are deployed, the achievable DoF region defined in (40) is reduced to

Definition 5.2. *(DoF region with RBF) The DoF region of a C-cell MIMO RBF system is defined as*

$$\mathcal{D}_{RBF\text{-}Rx} = \Big\{ (d_1, d_2, \cdots, d_C) \in \mathbb{R}_+^C : \forall(\omega_1, \omega_2, \cdots, \omega_C) \in \mathbb{R}_+^C;$$
$$\sum_{c=1}^{C} \omega_c d_c \leq \lim_{\rho \to \infty} \left[\max_{M_c \in \{0, \cdots, N_T\}} \sum_{c=1}^{C} \omega_c \frac{R_{RBF\text{-}Rx}^{(c)}}{\log_2 \rho} \right] \Big\}. \tag{41}$$

in which "Rx" denotes either MMSE, MF or AS.

Furthermore, the DoF regions are denoted as $\mathcal{D}_{MISO}$ and $\mathcal{D}_{RBF}$, respectively, under a multi-cell MISO setup. Certainly, $\mathcal{D}_{RBF} \subseteq \mathcal{D}_{MISO}$ and $\mathcal{D}_{RBF\text{-}Rx} \subseteq \mathcal{D}_{MIMO}$.

Note that the DoF region is, in general, applicable for any number of users per cell, K_c. However, if it is assumed that all K_c's are constant with $\rho \to \infty$, it can be shown that the DoF regions for the multi-cell RBF schemes given in (41) will collapse to the null point, i.e., a zero DoF for all the cells, due to the intra-/inter-cell interference[2]. It thus follows that for the analytical tractability, the DoF region characterization for the multi-cell RBF requires that K_c increases in a certain order with the SNR, ρ. [24] [22] thus make the following assumption[3]:

Assumption 1. *The number of users in each cell scales with ρ in the order of ρ^{α_c}, with $\alpha_c \geq 0$, denoted by $K_c = \Theta(\rho^{\alpha_c}), c = 1, \ldots, C$, i.e., $K_c/\rho^{\alpha_c} \to \delta$ as $\rho \to \infty$, with δ being a positive constant independent of α_c.*

---

[2] A rigorous proof of this claim can be deduced from Theorem 5.2 and 5.3 later for the special case of $\alpha_c = 0, \forall c$, i.e., all cells having a constant number of users.

[3] In [23], Tajer *et. al.* consider a more general assumption on the number of ad-hoc SISO links n, i.e., $\frac{\log n}{\log \rho} \to \alpha$ with $\alpha > 0$.

Considering the number of per-cell users to scale polynomially with the SNR is general as well as convenient. The linear scaling law, i.e., $K_c = \beta_c \rho$, with constant $\beta_c > 0$, is only a special case of $K_c = \Theta(\rho^{\alpha_c})$ with $\alpha_c = 1$; if K_c is a constant, then the corresponding α_c is zero. As will be shown later in this section, Assumption 1 enables us to obtain an efficient as well as insightful characterization of the DoF region for the multi-cell RBF. Note that the DoF region under Assumption 1 can be considered as a generalization of the conventional DoF region analysis based on IA [9] for the case of finite number of users, to the case of asymptotically large number of users that scales with the SNR. For the notational convenience, we use $\mathcal{D}_{\text{MISO}}(\boldsymbol{\alpha})$, $\mathcal{D}_{\text{MIMO}}(\boldsymbol{\alpha})$, $\mathcal{D}_{\text{RBF}}(\boldsymbol{\alpha})$, and $\mathcal{D}_{\text{RBF-Rx}}(\boldsymbol{\alpha})$ to denote the achievable DoF regions corresponding to $K_c = \Theta(\rho^{\alpha_c})$, $c = 1, \cdots, C$, and $\boldsymbol{\alpha} = [\alpha_1, \cdots, \alpha_C]^T$.

5.1. Multi-cell MISO RBF

In this subsection, we review the MISO RBF study in [22]. It is revealed that the high-SNR analysis provides an efficient way of characterizing the achievable sum rate for RBF, even for small values of SNR and number of users. In single-cell RBF, the DoF result shows that adjusting the number of beams according to the number of users can improve the achievable sum rate, thus providing a theoretical explanation for observations in, e.g., [46] [33] [48]. Extending to the multi-cell case, it is furthermore unfolded that collaboration among the BSs in assigning their respective numbers of data beams based on different per-cell user densities is essential to achieve the optimal throughput tradeoffs among different cells.

5.1.1. Single-cell case

First, we consider the DoF for the achievable sum rate in the single-cell RBF case without the ICI. The cell index c in (2) thus is dropped for brevity. In the single-cell case, the DoF region collapses to a line, bounded by 0 and $d^*_{\text{RBF}}(\alpha)$, where $d^*_{\text{RBF}}(\alpha) \geq 0$ denotes the maximum DoF achievable for the RBF sum rate.

The achievable DoF for single-cell RBF is defined with a given pair of α and M as

$$d_{\text{RBF}}(\alpha, M) = \lim_{\rho \to \infty} \frac{R_{\text{RBF}}}{\log_2 \rho} = \lim_{\eta \to \infty} \frac{R_{\text{RBF}}}{\log_2 \eta} \tag{42}$$

since $\eta = \rho/M$. Thus, $d^*_{\text{RBF}}(\alpha) = \max_{M \in \{1, \cdots, N_T\}} d_{\text{RBF}}(\alpha, M)$ for a given $\alpha \geq 0$. The DoF $d_{RBF}(\alpha, M)$ is characterized in the following lemma.

Lemma 5.1. *([22, Lemma 4.1]) Assuming $K = \Theta(\rho^\alpha)$, the DoF of single-cell MISO RBF with $M \leq N_T$ orthogonal transmit beams is given by*

$$d_{RBF}(\alpha, M) = \begin{cases} \dfrac{\alpha M}{M-1}, & 0 \leq \alpha \leq M-1, \quad \text{(43a)} \\ M, & \alpha > M-1. \quad \text{(43b)} \end{cases}$$

With RBF and under the assumption $K = \Theta(\rho^\alpha)$, it is interesting to observe from Lemma 5.1 that the achievable DoF can be a non-negative real number (as compared to the conventional integer DoF in the literature with a finite K). Moreover, it is observed that for any given $0 < \alpha < N_T - 1$, assigning more transmit beams by increasing M initially improves the sum-rate DoF if $M \leq \alpha + 1$; however, as $M > \alpha + 1$, the DoF may not necessarily increase with M due to the more dominant inter-beam/intra-cell interference. Note that the term $M - 1$ in the denominator of (43a) is exactly the number of interfering beams to one particular beam. Thus, Lemma 5.1 provides a succinct description of the interplay between the available multi-user diversity (specified by α with a larger α denoting a higher user density or the number of users in a cell), the level of the intra-cell interference (specified by $M - 1$), and the achievable spatial multiplexing gain or DoF, $d_{\text{RBF}}(\alpha, M)$.

It is also of great interest to obtain the maximum achievable DoF for a given α. The result is shown in the following theorem.

Theorem 5.1. *([22, Theorem 4.1]) For the single-cell MISO RBF with N_T transmit antennas and user density coefficient α, the maximum achievable DoF and the corresponding optimal number of transmit beams are*[4]

$$d^*_{RBF}(\alpha) = \begin{cases} \lfloor\alpha\rfloor + 1, & \alpha \leq N_T - 1, 1 \geq \{\alpha\}(\lfloor\alpha\rfloor + 2), \\ \frac{\alpha(\lfloor\alpha\rfloor+2)}{\lfloor\alpha\rfloor+1}, & \alpha \leq N_T - 1, \{\alpha\}(\lfloor\alpha\rfloor + 2) > 1, \\ N_T, & \alpha > N_T - 1. \end{cases} \tag{44}$$

$$M^*_{RBF}(\alpha) = \begin{cases} \lfloor\alpha\rfloor + 1, & \alpha \leq N_T - 1, 1 \geq \{\alpha\}(\lfloor\alpha\rfloor + 2), \\ \lfloor\alpha\rfloor + 2, & \alpha \leq N_T - 1, \{\alpha\}(\lfloor\alpha\rfloor + 2) > 1, \\ N_T, & \alpha > N_T - 1. \end{cases} \tag{45}$$

In Fig. 4, we compare the numerical sum rate and the analytical scaling law in Lemma 5.1. It is observed that the newly obtained sum-rate scaling law, $R_{\text{RBF}} \approx d_{\text{RBF}}(\alpha, M) \log_2 \rho$, in the single-cell RBF case is very accurate, even for small values of SNR ρ and number of users $K = \lfloor\rho^\alpha\rfloor$. Compared with Fig. 2 for the conventional scaling law, $R_{\text{RBF}} \approx M \log_2 \log K$, a much faster convergence is observed here. The DoF approach thus provides a more efficient way of characterizing the achievable sum rate for single-cell RBF. Also observe that the sum rate for $M = 2$ is higher than that for $M = 4$. This is because with $N_T = 4$ and $\alpha = 1$ in this example, the optimal number of beams to achieve $d^*_{RBF}(1) = 2$ is $M^*_{RBF}(1) = 2$ from (45). Since many previous studies have observed that adjusting the number of beams according to the number of users in single-cell RBF can improve the achievable sum rate (see, e.g., [46] [33] [48]), our study here can be considered as a theoretical explanation for such an observation.

Fig. 5 shows the maximum DoF and the corresponding optimal number of transmit beams versus the user density coefficient α with $N_T = 4$ for single-cell RBF, according to Theorem 5.1. It is observed that to maximize the achievable sum rate, we should only transmit more data beams when the number of users increases beyond a certain threshold. It is also observed that the maximum DoF $d^*_{RBF}(\alpha) = 4$ with $M = N_T = 4$ is attained when $\alpha \geq 3$ since $M^*_{RBF}(3) = 4$.

[4] The notations $\lfloor\alpha\rfloor$ and $\{\alpha\}$ denote the integer and fractional parts of a real number α, respectively.

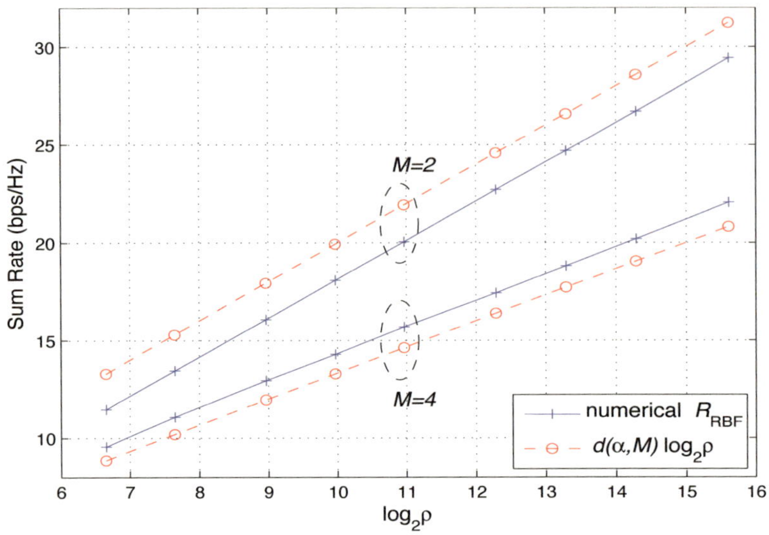

Figure 4. Comparison of the numerical sum rate and the scaling law $d_{RBF}(\alpha, M)\log_2\rho$, with $N_T = 4$, $\alpha = 1$, and $K = \lfloor \rho^{\alpha} \rfloor$.

5.1.2. Multi-cell case

We then extend the high-SNR analysis for the single-cell RBF to the more general multi-cell RBF subject to the ICI. For convenience, we denote the achievable sum-rate DoF of the c-th cell as $d_{\text{RBF},c}(\alpha_c, \boldsymbol{m}) = \lim_{\rho\to\infty} \frac{R_{\text{RBF}}^{(c)}}{\log_2 \rho}$, where $\boldsymbol{m} = [M_1, \cdots, M_C]^T$ is a given set of numbers of transmit beams at different BSs. We then state the next lemma on the achieve DoF of the c-th cell.

Lemma 5.2. *([22, Lemma 4.2]) In the multi-cell MISO RBF, assuming $K_c = \Theta(\rho^{\alpha_c})$, the achievable DoF of the c-th cell $d_{\text{RBF},c}(\alpha_c, \boldsymbol{m})$, $c \in \{1, \ldots, C\}$, for a given $\boldsymbol{m}$ is*

$$d_{RBF,c}(\alpha_c, \boldsymbol{m}) = \begin{cases} \dfrac{\alpha_c M_c}{\sum_{l=1}^{C} M_l - 1}, & 0 \le \alpha_c \le \sum_{l=1}^{C} M_l - 1, \quad (46a) \\ M_c, & \alpha_c > \sum_{l=1}^{C} M_l - 1. \quad (46b) \end{cases}$$

Similar to Lemma 5.1 for the single-cell case, Lemma 5.2 reveals the relationship among the multi-user diversity, the level of the interference, and the achievable DoF for multi-cell RBF. However, as compared to the single-cell case, there are not only $M_c - 1$ intra-cell interfering beams, but also $\sum_{l=1, l\neq c}^{C} M_l$ inter-cell interfering beams for any beam of the c-th cell in the multi-cell case, as observed from the denominator in (46a), which results in a decrease in the achievable DoF per cell.

Next, we obtain characterization of the DoF region defined in (41) for the multi-cell RBF with any given set of per-cell user density coefficients, denoted by $\boldsymbol{\alpha} = [\alpha_1, \cdots, \alpha_C]^T$ in the following theorem; for convenience, we denote $\boldsymbol{d}_{\text{RBF}}(\boldsymbol{\alpha}, \boldsymbol{m}) = \left[\, d_{\text{RBF},1}(\alpha_1, \boldsymbol{m}) \,, \cdots, d_{RBF,C}(\alpha_C, \boldsymbol{m})\right]^T$, with $d_{\text{RBF},c}(\alpha_c, \boldsymbol{m})$ given in Lemma 5.2.

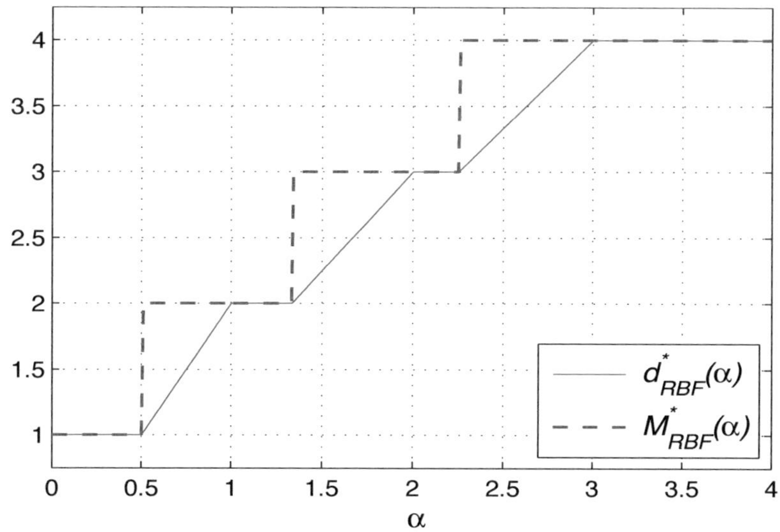

Figure 5. The maximum DoF $d^*_{RBF}(\alpha)$ and optimal number of beams $M^*_{RBF}(\alpha)$ with $N_T = 4$.

Theorem 5.2. *([22, Theorem 4.2]) Assuming $K_c = \Theta\left(\rho^{\alpha_c}\right), c = 1, \ldots, C$, the achievable DoF region of a C-cell RBF system is given by*

$$\mathcal{D}_{RBF}(\boldsymbol{\alpha}) = \boldsymbol{conv}\Big\{ \boldsymbol{d}_{RBF}(\boldsymbol{\alpha}, \boldsymbol{m}), M_c \in \{0, \cdots, N_T\},\ c = 1, \cdots, C \Big\}, \tag{47}$$

where ***conv*** *denotes the convex hull operation.*

This theorem implies that we can obtain the DoF region of multi-cell RBF $\mathcal{D}_{\text{RBF}}(\boldsymbol{\alpha})$ by taking the convex hull over all achievable DoF points $\boldsymbol{d}_{\text{RBF}}(\boldsymbol{\alpha}, \boldsymbol{m})$ with all different values of $\boldsymbol{m}$, i.e., different BS beam number assignments.

In Fig. 6, we depict the DoF region of a two-cell RBF system with $N_T = 4$, and for different user density coefficients α_1 and α_2. The vertices of these regions can be obtained by setting appropriate numbers of beams $0 \leq M_1 \leq 4$ and $0 \leq M_2 \leq 4$, while time-sharing between these vertices yields the entire boundary. To achieve the maximum sum-DoF of both cells, it is observed that a rule of thumb is to transmit more beams in the cell with a higher user density, and when α_1 and α_2 are both small, even turn off the BS of the cell with the smaller user density. Since the maximum sum-DoF does not consider the throughput fairness, the DoF region clearly shows all the achievable sum-rate tradeoffs among different cells, by observing its (Pareto) boundary as shown in Fig. 6. It is also observed that switching the two BSs to be on/off alternately achieves the optimal DoF boundary when the numbers of users in both cells are small, but is strictly suboptimal when the user number becomes large (see the dashed line in Fig. 6).

Furthermore, consider the case without any cooperation between these two BSs in assigning their numbers of transmit beams, i.e., both cells act selfishly by transmitting $M_c = N_T$ beams to intend to maximize their own DoF. The resulted DoF pairs $\boldsymbol{d}_{\text{RBF}}([\alpha_1, \alpha_2], [4, 4])$ for three

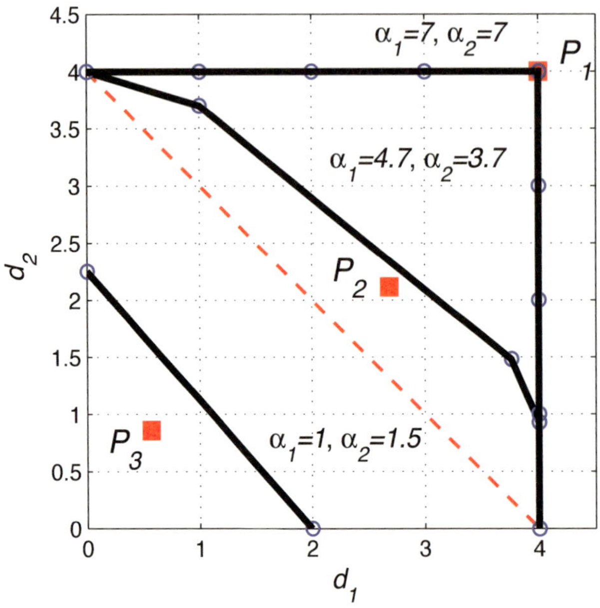

Figure 6. DoF region of two-cell RBF system with $N_T = 4$.

sets of $\boldsymbol{\alpha}$ are shown in Fig. 6 as P_1, P_2, and P_3, respectively. It is observed that the smaller the user densities are, the further the above non-cooperative multi-cell RBF scheme deviates from the Pareto boundary. In general, the optimal DoF tradeoffs or the boundary DoF pairs are achieved when both cells cooperatively assign their numbers of transmit beams based on their respective user densities, especially when the numbers of users in both cells are not sufficiently large. Since the information needed to determine the optimal operating DoF point is only the individual cell user density coefficients, the DoF region provides a very useful method to globally optimize the coordinated multi-cell RBF scheme in practical systems.

5.2. Multi-cell MIMO RBF schemes: Benefit of spatial receive diversity

Spatial receive diversity is another topic which is not fully understood even in a single-cell RBF setup. As discussed in Section 4.2, RBF with and without receive diversity schemes still achieve the same sum-rate scaling law, assuming that the number of users per cell goes to infinity for any given user's SNR. Based on the conventional asymptotic analysis, it is thus concluded that receive diversity only provides *marginal* gain to the rate performance.

The objective of this subsection is to review the recent results on the rate performance of multi-cell MIMO RBF systems under the high-SNR/DoF analysis [24]. Contrasting with the large-number-of-users analysis, it is found that the benefit of receive diversity is significant for the RBF scheme. In particular, receive diversity introduces an interference mitigation effect on the DoF performance. As a consequence, an substantial less number of users in each cell is required to achieve a given DoF region comparing to the case without receive diversity. However, the gain is observed only with MMSE receiver but not with suboptimal schemes such as MF or AS. This reflects a tradeoff between the rate/DoF performance and the complexity/delay time of RBF systems.

5.2.1. DoF of the c-th cell

Similarly to the MISO case, the DoF of the c-th RBF-Rx cell is defined as $d_{\text{RBF-Rx},c}(\alpha_c, \boldsymbol{m}) = \lim_{\rho\to\infty} \frac{R^{(c)}_{\text{RBF-Rx}}}{\log_2 \rho}$. We then obtain the DoF of the c-th cell.

Lemma 5.3. *([24, Lemma 4.1]) In the multi-cell RBF-MMSE, assuming $K_c = \Theta(\rho^{\alpha_c})$, the achievable DoF of the c-th cell $d_{RBF\text{-}MMSE,c}(\alpha_c, \boldsymbol{m})$, $c \in \{1, \ldots, C\}$, for a given $\boldsymbol{m}$ is*

$$d_{RBF\text{-}MMSE,c}(\alpha_c, \boldsymbol{m}) = \begin{cases} \dfrac{\alpha_c M_c}{\sum_{l=1}^{C} M_l - N_R}, & 0 \le \alpha_c \le \sum_{l=1}^{C} M_l - N_R, \quad (48a) \\ M_c, & \alpha_c > \sum_{l=1}^{C} M_l - N_R. \quad (48b) \end{cases}$$

Lemma 5.4. *([24, Lemma 4.2]) In the multi-cell RBF-MF/AS, assuming $K_c = \Theta(\rho^{\alpha_c})$, the achievable DoF of the c-th cell $d_{RBF\text{-}MF/AS,c}(\alpha_c, \boldsymbol{m})$, $c \in \{1, \ldots, C\}$, for a given $\boldsymbol{m}$ is*

$$d_{RBF\text{-}MF/AS,c}(\alpha_c, \boldsymbol{m}) = \begin{cases} \dfrac{\alpha_c M_c}{\sum_{l=1}^{C} M_l - 1}, & 0 \le \alpha_c \le \sum_{l=1}^{C} M_l - 1, \quad (49a) \\ M_c, & \alpha_c > \sum_{l=1}^{C} M_l - 1. \quad (49b) \end{cases}$$

The DoF scaling laws $d_{\text{RBF-MMSE},c}(\alpha_c, \boldsymbol{m}) \log_2 \rho$ and $d_{\text{RBF-MF/AS},c}(\alpha_c, \boldsymbol{m}) \log_2 \rho$ are compared in Fig. 7. As a reference, we also show the sum-rates of the RBF-MMSE, RBF-MF, and RBF-AS. Again, a good match is observed between the scaling law and the numerical results, even with small values of the SNR and number of users.

We first remark that the MMSE receiver produces an *interference mitigation* effect to the DoF performance. From Lemma 5.2 and 5.3, it is observed that $N_R - 1$ interferences can be effectively eliminated in the RBF-MMSE. In particular, with N_R degrees of freedom at the receiver, the users can use one for receiving signal and $N_R - 1$ for suppressing the interferences. Secondly, in terms of sum-rate, the gain captured by employing either RBF-MF or RBF-AS is only marginal comparing to the MISO RBF with single-antenna users, as shown in Lemma 5.4. The benefit of receive diversity, therefore, clearly depends on the availability of the interferences' CSI at the users. In the RBF-MMSE, this CSI is the interference-plus-noise covariance matrix $\boldsymbol{W}_k^{(c)}$. In fact, since there is no information about the interferences given to the users in the RBF-MF/AS, it is expected that no significant gain is achieved by these schemes comparing with the MISO RBF.

At this point, we observe a sharp contrast between the conclusions based on the high-SNR/DoF and large-number-of-users analysis. In the conventional analysis, Proposition 4.2 effectively implies that there is no significant rate gain for the multi-cell RBF with receive diversity scheme at the users, and also no loss with intra-/inter-cell interferences in the system as the number of users goes to infinity. Recalling the convergences of the two approaches, the high-SNR/DoF analysis, therefore, is a more efficient method

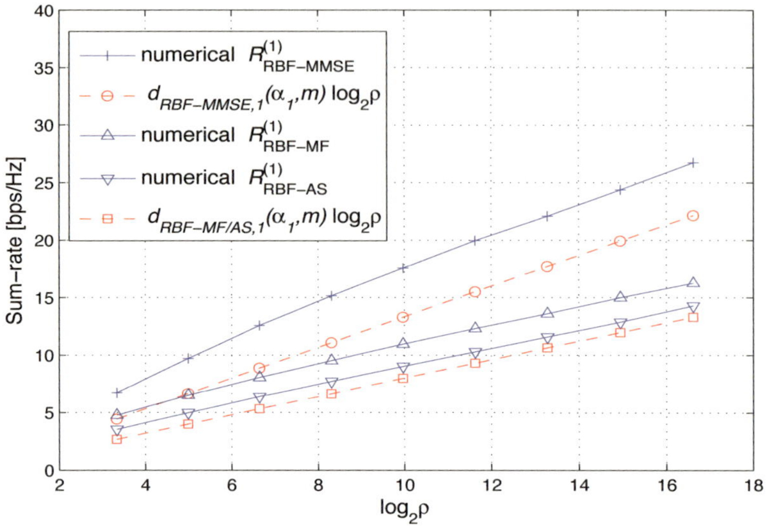

Figure 7. Comparison of the numerical sum-rate and the DoF scaling law, with $C = 2$, $[M_1, M_2] = [4, 2]$, $N_R = 3$, $\alpha_1 = 1$, $\gamma_{2,1} = -1$dB, and $K_1 = \lfloor \rho^{\alpha_1} \rfloor$.

of characterizing the achievable sum-rate for RBF under the effects of multi-user diversity, spatial receive diversity, and interference.

5.2.2. DoF region characterization

We denote $\boldsymbol{d}_{\text{RBF-Rx}}(\boldsymbol{\alpha}, \boldsymbol{m}) = \left[\, d_{\text{RBF-Rx},1}(\alpha_1, \boldsymbol{m}) \;, \cdots, \; d_{\text{RBF-Rx},C}(\alpha_C, \boldsymbol{m})\right]^T$, with $d_{\text{RBF-Rx},c}(\alpha_c, \boldsymbol{m})$ given in Lemma 5.3 or 5.4. The characterization of the DoF region for the multi-cell RBF-Rx is given in the following theorem

Theorem 5.3. *Assuming $K_c = \Theta\left(\rho^{\alpha_c}\right), c = 1, \ldots, C$, the achievable DoF region of a C-cell RBF-Rx system is given by*

$$\mathcal{D}_{\text{RBF-Rx}}(\boldsymbol{\alpha}) = \textbf{conv}\Big\{ \boldsymbol{d}_{\text{RBF-Rx}}(\boldsymbol{\alpha}, \boldsymbol{m}), M_c \in \{0, \cdots, N_T\}, \; c = 1, \cdots, C \Big\}, \tag{50}$$

*where **conv** denotes the convex hull operation and "Rx" stands for either MMSE, MF, or AS as usual.*

Fig. 8 depicts the DoF region of a two-cell system employing either the RBF-MMSE or RBF-MF/AS. We assume that $N_T = 4$, $N_R = 2$. The region's boundaries for RBF-MMSE and RBF-MF/AS are denoted by solid and dashed lines, respectively. When α_1 and α_2 are small, the DoF region is greatly expanded in the case of the RBF-MMSE. In fact, compared to the RBF-MF, RBF-AS, and MISO RBF, an *exponentially* less number of users in each cell is required to achieve a certain DoF region in the RBF-MMSE (see Lemma 5.3). The assistance of receive diversity to multi-user diversity, therefore, is significant.

We see that receive diversity is indeed beneficial for RBF systems. However, there exists a tradeoff between the rate/DoF performance and the complexity/delay time. The option thus

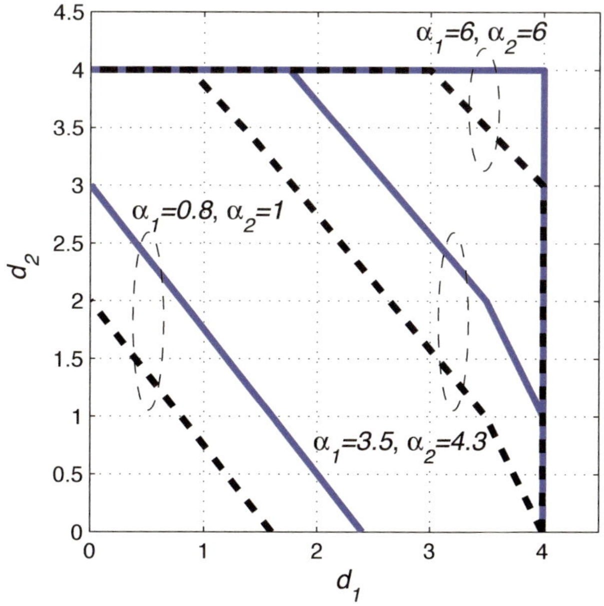

Figure 8. DoF regions of different RBF-MMSE and RBF-MF/AS systems.

depends on the communication system in consideration. The RBF-MMSE is a good choice if there are small numbers of users in the cells while the constrains on the complexity and delay time are slack. However, when the receivers are required to be simple and there are many users in the system, the RBF-MF, RBF-AS, or even MISO RBF, is more favourable.

5.3. Optimality of multi-cell RBF

So far, we have characterized the achievable DoF region for the multi-cell RBF schemes that require only partial CSI at the transmitter. One important question that remains unaddressed yet is how the multi-cell RBF performs as compared to the optimal transmission scheme (e.g., IA) for the multi-cell downlink system with the full transmitter CSI, in terms of achievable DoF region. In this subsection, we attempt to partially answer this question by focusing on the high-SNR/DoF regime. Note that we only consider the multi-cell MIMO RBF schemes. Discussions for the MISO RBF can be drawn from either the RBF-MMSE, RBF-MF, or RBF-AS by setting $N_R = 1$.

5.3.1. Single-cell case

First, we consider the single-cell case to draw some useful insights. It is well known that the maximum sum-rate DoF for a single-cell MIMO-BC with N_T transmit antennas and $K \geq N_T$ users each with N_R receive antennas under independent channels is N_T [1] [55], which is achievable by the DPC scheme or even simple linear precoding schemes. However, it is not immediately clear whether such a result still holds for the case of $K = \Theta(\rho^{\alpha}) \gg N_T$ with $\alpha > 0$, since in this case N_T may be only a lower bound on the maximum DoF. We thus have the following proposition.

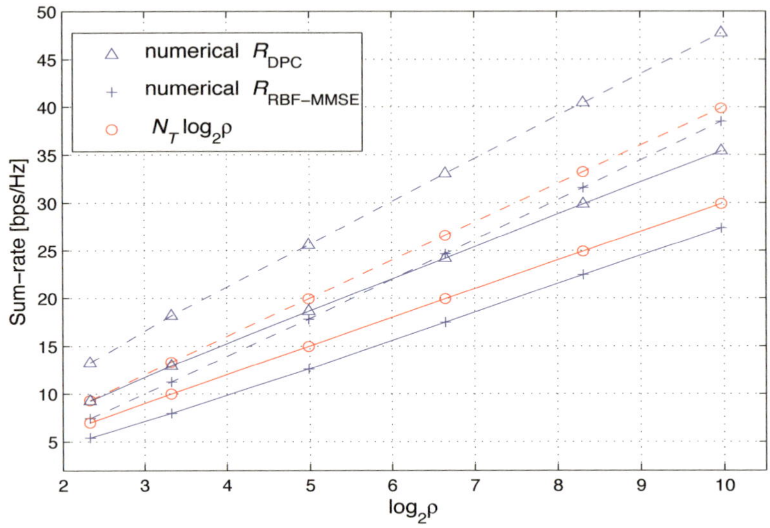

Figure 9. Comparison of the numerical DPC and RBF-MMSE sum-rates, and the DoF scaling law with $\alpha \geq N_T - N_R$.

Proposition 5.1. *([24, Proposition 4.1]) Assuming* $K = \Theta(\rho^\alpha)$ *with* $\alpha > 0$*, the maximum sum-rate DoF of a single-cell MIMO-BC with* N_T *transmit antennas is* $d^*(\alpha) = N_T$*.*

Proposition 5.1 confirms that the maximum DoF of the MIMO-BC is still N_T, even with the asymptotically large number of users that scales with SNR, i.e., multi-user diversity does not yield any increment of "interference-free" DoF. The optimality of the single-cell RBF schemes is given in the following proposition

Proposition 5.2. *([24, Proposition 4.4]) Assuming* $K = \Theta(\rho^\alpha)$*, the single-cell MIMO RBF-Rx scheme is DoF-optimal if*

- *RBF-MMSE:* $\alpha \geq N_T - N_R$.
- *RBF-MF/AS:* $\alpha \geq N_T - 1$.

From the above proposition, it follows that the single-cell MIMO RBF schemes achieve the maximum DoF with $M = N_T$ if the number of users is sufficiently large, thanks to the multi-user diversity effect that completely eliminates the intra-cell interference with a large number of users.

As an example, we compare the numerical sum-rates and the DoF scaling law in Fig. 9, in which the DPC and RBF-MMSE are employed, and $\alpha \geq N_T - N_R$. We consider two single-cell systems with the following parameters: (1) $M = N_T = 3$, $N_R = 2$, $\alpha = 1$, $K = \lfloor \rho^\alpha \rfloor$; and (2) $M = N_T = 4$, $N_R = 3$, $\alpha = 1.2$, $K = \lfloor \rho^\alpha \rfloor$. The rates and scaling law of system 1 and 2 are denoted as the solid and dash lines, respectively. Note that in both cases, the DPC and RBF-MMSE sum-rates follow the (same) DoF scaling law quite closely. This example clearly demonstrates the optimality of the RBF given a large per-cell number of users.

5.3.2. Multi-cell case

For the convenience of analysis, we use $\mathcal{D}_{\text{UB}}(\boldsymbol{\alpha})$ to denote an upper bound on the DoF region defined in (40), for a given $\boldsymbol{\alpha}$ in the multi-cell case. Clearly, under Assumption 1, it follows that $\mathcal{D}_{\text{RBF-Rx}}(\boldsymbol{\alpha}) \subseteq \mathcal{D}_{\text{MIMO}}(\boldsymbol{\alpha}) \subseteq \mathcal{D}_{\text{UB}}(\boldsymbol{\alpha})$.

The following proposition establishes the DoF region upper bound $\mathcal{D}_{\text{UB}}(\boldsymbol{\alpha})$.

Proposition 5.3. *([24, Proposition 4.3]) Given $K_c = \Theta(\rho^{\alpha_c})$, $c = 1, \cdots, C$, a DoF region upper bound for a C-cell downlink system is given by*

$$\mathcal{D}_{\text{UB}}(\boldsymbol{\alpha}) = \left\{ (d_1, d_2, \cdots d_C) \in \mathbb{R}_+^C : d_c \leq N_T, c = 1, \cdots, C \right\}. \tag{51}$$

The optimality of the multi-cell MIMO RBF-Rx schemes is shown in the following proposition

Proposition 5.4. *([24, Proposition 4.3]) Given $K_c = \Theta(\rho^{\alpha_c})$, $c = 1, \cdots, C$, the multi-cell MIMO RBF-Rx scheme achieves the "interference-free" DoF region of a C-cell downlink system, i.e., $\mathcal{D}_{RBF\text{-}Rx}(\boldsymbol{\alpha}) \equiv \mathcal{D}_{\text{UB}}(\boldsymbol{\alpha})$, if*

- *RBF-MMSE: $\alpha_c \geq CN_T - N_R$, $\forall c \in \{1, \cdots, C\}$.*
- *RBF-MF/AS: $\alpha_c \geq CN_T - 1$, $\forall c \in \{1, \cdots, C\}$.*

As a consequence, we also have $\mathcal{D}_{RBF\text{-}Rx}(\boldsymbol{\alpha}) \equiv \mathcal{D}_{MIMO}(\boldsymbol{\alpha})$, i.e., the RBF-Rx scheme is optimal given large per-cell user densities.

The above proposition implies that the multi-cell MIMO RBF schemes are indeed DoF-optimal when the numbers of users in all cells are sufficiently large. Due to the overwhelming multi-user diversity gain, RBF compensates the lack of full CSI at transmitters without any compromise of DoF degradation. However, it is important to point out that such a result should not undermine the benefits of having the more complete CSI at transmitters in practical multi-cell systems, where more sophisticated precoding schemes than RBF such as IA-based ones can be applied to achieve substantial throughput gains, especially when the numbers of per-cell users are not so large.

6. Summary

In this chapter, we have introduced the OBF/RBF and the current developments in the literature. First of all, we have given an overview for the single-cell case, summarizing some of the most important contributions so far. Furthermore, we have reviewed the recent investigations on the rate performance of multi-cell RBF systems in both finite- and high-SNR regimes. These results are useful for the optimal design of multi-cell RBF in practical cellular systems. In particular, it is revealed that collaboration among the BSs in assigning their respective numbers of data beams based on different per-cell user densities is essential to achieve the optimal throughput tradeoffs among different cells. Moreover, the results show that spatial receive diversity is also a significant factor to be considered, noting that there exists, however, a tradeoff between the rate/DoF performance and the complexity/delay time of RBF systems with different receivers. The preference of multi-cell RBF is justified by the scheme's optimality albeit requiring only partial CSI at transmitters as compared to other full-CSI transmission schemes, when the numbers of users in all cells are sufficiently large.

Author details

Hieu Duy Nguyen, Rui Zhang, and Hon Tat Hui

* Address all correspondence to: nguyendh@i2r.a-star.edu.sg, elezhang@nus.edu.sg, elehht@nus.edu.sg

Department of Electrical and Computer Engineering, National University of Singapore, Singapore

References

[1] G Caire and S Shamai. On the achievalbe throughput of a multi-antenna Gaussian broadcast channel. *IEEE Trans. Inf. Theory*, 49:1691–1706, July 2003.

[2] M Costa. Writing on dirty paper. *IEEE Trans. Inf. Theory*, 29:439–441, May 1983.

[3] H Weingarten, Y Steinberg, and S Shamai. The capacity region of the Gaussian multiple-input multiple-output broadcast channel. *IEEE Trans. Inf. Theory*, 52:3936–3964, Sep. 2006.

[4] Q H Spencer, A L Swindlehurst, and M Haardt. Zero-forcing methods for downlink spatial multiplexing in multiuser MIMO channels. *IEEE Trans. Signal Process.*, 52:461–471, Feb. 2004.

[5] E Biglieri, A Constantinides, R Calderbank, A Goldsmith, A Paulraj, and V Poor. *Introduction to MIMO Wireless Communications*. Cambridge Univ. Press, June 2006.

[6] A B Gershman and N Sidiropoulos, editors. *Space-Time Processing for MIMO Communications*. John Wiley & Sons, 2005.

[7] A Paulraj, R Nabar, and D Gore. *Introduction to Space-Time Wireless Communications*. Cambridge Univ. Press, May 2003.

[8] A E Gamal and Y-H Kim. *Network Information Theory*. Cambridge University Press, 2011.

[9] S A Jafar. *Interference alignment: a new look at signal dimensions in a communication network*. Foundations and Trends in Communications and Information Theory, 2011.

[10] B Ng, J Evans, S Hanly, and D Aktas. Distributed downlink beamforming with cooperative base stations. *IEEE Trans. Inf. Theory*, 54:5491–5499, Dec. 2008.

[11] L Zhang, R Zhang, Y C Liang, Y Xin, and H V Poor. On the Gaussian MIMO BC-MAC duality with multiple transmit covariance constraints. *IEEE Trans. Inf. Theory*, 58:2064–2078, Apr. 2012.

[12] R Zhang. Cooperative multi-cell block diagonalization with per-base-station power constraints. *IEEE J. Sel. Areas Commun.*, 28:1435–1445, Dec. 2010.

[13] H Dahrouj and W Yu. Coordinated beamforming for the multicell multi-antenna wireless systems. *IEEE Trans. Wireless Commun.*, 9:1748–1795, May 2010.

[14] Y-F Liu, Y-H Dai, and Z-Q Luo. Coordinated beamforming for MISO interference channel: complexity analysis and efficient algorithms. *IEEE Trans. Signal Process.*, 59:1142–1157, Mar. 2011.

[15] E Bjornson, R Zakhour, D Gesbert, and B Ottersten. Cooperative multicell precoding: rate region characterization and distributed strategies with instantaneous and statistical CSI. *IEEE Trans. Signal Process.*, 58:4298–4310, Aug. 2010.

[16] X Shang, B Chen, and H V Poor. Multiuser MISO interference channels with single-user detection: optimality of beamforming and the achievable rate region. *IEEE Trans. Inf. Theory*, 57:4255–4273, July 2011.

[17] R Zhang and S Cui. Cooperative interference management with MISO beamforming. *IEEE Trans. Signal Process.*, 58:5450–5458, Oct. 2010.

[18] N Jindal. MIMO broadcast channels with finite-rate feedback. *IEEE Trans. Inf. Theory*, 52:5045–5060, Nov. 2006.

[19] P Viswanath, D N C Tse, and R Laroia. Opportunistic beamforming using dumb antennas. *IEEE Trans. Inf. Theory,*, 48:1277–1294, June 2002.

[20] M Sharif and B Hassibi. On the capacity of MIMO broadcast channel with partial side information. *IEEE Trans. Inf. Theory*, 51:506–522, Feb. 2005.

[21] M Sharif and B Hassibi. A comparison of time-sharing, beamforming, and DPC for MIMO broadcast channels with many users. *IEEE Trans. Commun.*, 55:11–15, Jan. 2007.

[22] H D Nguyen, R Zhang, and H T Hui. Multi-cell random beamforming: achievable rate and degree of freedom region. *submitted to IEEE Trans. Sig. Proc., available online at http://arxiv.org/abs/1205.5849*, 2012.

[23] A Tajer and X Wang. (n, K)-user interference channel: degrees of freedom. *IEEE Trans. Inf. Theory*, 58:5338–5353, Aug. 2012.

[24] H D Nguyen, R Zhang, and H T Hui. Effect of spatial receive diversity on the degree of freedom region of multi-cell random beamforming. *in preparation*, 2012.

[25] H-C Yang, P Lu, H-K Sung, and Y-C Ko. Exact sum-rate analysis of MIMO broadcast channels with random unitary beamforming. *IEEE Trans. Commun.*, 59:2982–2986, Nov. 2011.

[26] Y-C Ko, H-C Yang, S-S Eom, and M-S Alouini. Adaptive modulation with diversity combining based on output-threshold MRC. *IEEE Trans. Wireless Commun.*, 6:3728–3737, Oct. 2007.

[27] Y Kim, J Yang, and D K Kim. A closed form approximation of the sum-rate upper bound of random beamforming. *IEEE Commun. Lett.*, 12:365–367, May 2008.

[28] D Park and S Y Park. Performance analysis of multiuser diversity under transmit antenna correlation. *IEEE Trans. Commun.*, 56:666–674, Apr. 2008.

[29] H David and H Nagaraja. *Order Statistics*. New York: Wiley, 3rd edition, 2003.

[30] K Huang, J G Andrews, and R W Heath, Jr. Performance of orthogonal beamforming for SDMA with limited feedback. *IEEE Trans. Veh. Tech.*, 58:152–164, Jan. 2009.

[31] K Huang, R W Heath, and J G Andrews. Space division multiple access with a sum feedback rate constraint. *IEEE Trans. Veh. Tech.*, 58:3879–3891, Jul. 2007.

[32] L Sun and M R McKay. Eigen-based transceivers for the MIMO broadcast channel with semi-orthogonal user selection. *IEEE Trans. Sig. Proc.*, 58:5246–5261, Oct. 2010.

[33] J L Vicario, R Bosisio, C Anton-Haro, and U Spagnolini. Beam selection strategies for orthogonal random beamforming in sparse networks. *IEEE Trans. Wireless Commun.*, 7:3385–3396, Sep. 2008.

[34] T Yoo and A Goldsmith. On the optimality of multi-antenna broadcast scheduling using zero-forcing beamforming. *IEEE J. Sel. Areas Commun.*, 24:528–541, Mar. 2006.

[35] R Couillet, J Hoydis, and M Debbah. Random beamforming over quasi-static and fading channels: a deterministic equivalent approach. *to appear in IEEE Trans. Inf. Theory, available online at http://arxiv.org/abs/1011.3717*, 2012.

[36] O Ozdemir and M Torlak. Optimum feedback quantization for an opportunistic beamforming scheme. *IEEE Trans. Wireless Commun.*, 9:1584–1593, May 2010.

[37] S Sanayei and A Nosratinia. Opportunistic beamforming with limited feedback. *IEEE Trans. Wireless Commun.*, 6:2765–2770, Aug. 2007.

[38] Y Xue and T Kaiser. Exploiting multiuser diversity with imperfect one-bit channel state feedback. *IEEE Trans. Veh. Tech.*, 56:183–193, May 2007.

[39] J So and J M Cioffi. Feedback reduction scheme for downlink multi-user diversity. *IEEE Trans. Wireless Commun.*, 8:668–672, Feb. 2009.

[40] A Rajanna and N Jindal. Multi-user diversity in downlink channels: when does the feedback cost outweigh the spectral efficiency benefit. *IEEE Trans. Wireless Commun.*, 11:408–418, Jan. 2012.

[41] A Barg and D Yu Nogin. Bounds on packings of spheres in the Grassmann manifold. *IEEE Trans. Inf. Theory*, 48:2450–2454, Sep. 2002.

[42] B M Hochwald, T L Marzetta, T J Richardson, W Sweldens, and R Urbanke. Systematic design of unitary space-time constellations. *IEEE Trans. Inf. Theory*, 46:1962–1973, Sep. 2000.

[43] D J Love and R W Heath. Grassmannian beamforming for multiple-input multiple-ouput wireless systems. *IEEE Trans. Inf. Theory*, 49:2735–2747, Oct. 2003.

[44] N Zorba and A I Perez-Neira. Opportunistic Grassmannian beamforming for multi-user and multi-antenna downlink communications. *IEEE Trans. Wireless Commun.*, 7:1174–1178, Apr. 2008.

[45] M Xia, Y-C Wu, and S Aissa. Non-orthogonal opportunistic beamforming: performance analysis and implementation. *IEEE Trans. Wireless Commun.*, 11:1424–1433, Apr. 2012.

[46] M Kountouris, D Gesbert, and T Salzer. Enhanced multiuser random beamforming: dealing with the not-so-large number of users case. *IEEE Trans. Wireless Commun.*, 26:1536–1545, Oct. 2008.

[47] H Kwon, E W Jang, and J M Cioffi. Predetermined power allocation for opportunistic beamforming with limited feedback. *IEEE Trans. Wireless Commun.*, 10:84–90, Jan. 2011.

[48] J Wagner, Y C Liang, and R Zhang. On the balance of multiuser diversity and spatial multiplexing gain in random beamforming. *IEEE Trans. Wireless Commun.*, 7:2512–2525, July 2008.

[49] I H Kim, S Y Park, D J Love, and S J Kim. Improved multi-user MIMO unitary precoding using partial channel state information and insights from the Reimannian manifold. *IEEE Trans. Wireless Commun.*, 8:4014–4022, Aug. 2009.

[50] H Kushner and P A Whiting. Convergence of proportional-fair sharing algorithms under general conditions. *IEEE Trans. Wireless Commun.*, 3:1250–1259, July 2004.

[51] H Zhou, P Fan, and J Li. Global proportional fair scheduling for networks with multiple base stations. *IEEE Trans. Veh. Tech.*, 60:1867–1879, May 2011.

[52] I S Gradshteyn and I M Ryzhik. *Table of Integrals, Series and Products*. Jeffrey, A. & Zwillinger, D. (eds), Academic Press, 7th edition, 2007.

[53] S-H Moon, S-R Lee, and I Lee. Sum-rate capacity of random beamforming for multi-antenna broadcast channels with other cell interference. *IEEE Trans. Wireless Commun.*, 10:2440–2444, Aug. 2011.

[54] H Gao and P J Smith. Exact SINR calculations for optimum linear combining in wireless systems. *Prob. Eng. Inf. Sci.*, 40:261–281, May 1998.

[55] Y Huang and B D Rao. Closed form sum-rate of random beamforming. *IEEE Commun. Lett.*, 16: 630-633, May 2012.

Section 3

Multi-User MIMO in LTE Precoding Design

Chapter 3

Multi-user MIMO in LTE and LTE-Advanced - Receiver Structure and Precoding Design

Rizwan Ghaffar, Raymond Knopp and
Florian Kaltenberger

Additional information is available at the end of the chapter

http://dx.doi.org/10.5772/57134

1. Introduction

The spatial dimension surfacing from the usage of multiple antennas promises improved reliability, higher spectral efficiency [24], and the spatial separation of users [6]. This spatial dimension (MIMO) is particularly beneficial for precoding in the downlink of multi-user cellular systems (broadcast channel), where these spatial degrees of freedom at the transmitter can be used to transmit data to multiple users simultaneously. This is achieved by creating independent parallel channels to the users (canceling multi-user interference) and the users subsequently employ simplified single-user receiver structures. However, the transformation of cross-coupled channels into parallel non-interacting channels necessitates perfect channel state information at the transmitter (CSIT) whose acquisition in a practical system, in particular frequency division duplex (FDD) system, is far from realizable. The complexity associated with the feedback overhead coupled with the low rate feedback channels are the major impediments in CSIT acquisition. This leads to the precoding strategies based on the partial or quantized CSIT [15], which limit the gains of multi-user MIMO.

On the design of feedback, there is stark contrast between the theoretically established results and current standards. Theory has established that the amount of CSIT feedback in a downlink system needs to grow in proportion to the SNR [11] and otherwise the degrees of freedom are lost. However to avoid the burden of feedback and due to complexity constraints, the modern wireless systems have been restricted to fixed rate feedback schemes. With such premises, LTE and LTE-Advanced have focused on the structured precoder codebook based approach [17, 19] by using a small number of feedback bits. These LTE precoders are characterized by low-resolution and are further based on the principle of equal gain transmission (EGT). These precoders when employed for the multi-user MIMO mode of transmission are unable to cancel the multi-user interference thereby increasing the sub-optimality of conventional single-user detection. This fixed low-level quantization of LTE codebook, therefore, eclipses most of the benefits of multi-user MIMO and raises questions about the feasibility of this mode of transmission [22, page 244]. This strong perception is based on the fact that users can not cooperate in multi-user scenario and further on the assumption that users employ simple single-user receivers.

In this chapter, we focus on a new paradigm of multi-user MIMO where users exploit the discrete structure of interference, instead of ignoring it or assuming it to be Gaussian and merging it in noise. We compare the two strategies of interference exploitation and interference cancellation in multi-user scenario. For the former, we look at low complexity multi-user detectors. Though multi-user detection has been extensively investigated in the literature for the uplink (multiple access channel), its related complexity has so far prohibited its employment in the downlink (broadcast channel). For the multiple access channel, several multi-user detection techniques exist in the literature starting from the optimal multi-user receivers [25] to their near-optimal reduced complexity counterparts (sphere decoders [3]). The complexity associated with these techniques led to the investigation of low-complexity solutions as sub-optimal linear multi-user receivers [20], iterative multi-user receivers [26, 28], and decision-feedback receivers [5, 12]. Since in practice, most wireless systems employ error control coding combined with the interleaving , recent work in this area has addressed multi-user detection for coded systems based on soft decisions [13, 23]. We focus in this chapter on a low-complexity interference-aware receiver structure which not only reduces one complex dimension of the system but is also characterized by exploiting the interference structure in the detection process. Considering this receiver structure, we investigate the effectiveness of the low-resolution LTE precoders for the multi-user MIMO mode and show that multi-user MIMO can bring significant gains in future wireless systems if the users resort to intelligent interference-aware detection as compared to the sub-optimal single-user detection.

In an effort of bridging the gap between the theoretical and practical gains of multi-user MIMO, this chapter investigates the structure of LTE codebook by analyzing the pairwise error probability (PEP) expressions. The analysis shows that LTE precoders suffer from the loss of diversity when being employed in multi-user MIMO transmission mode but no such loss is observed in single-user MIMO mode. Based on this analysis, a new codebook design is proposed and it is shown that with a nominal increase in the feedback, the performance of multi-user MIMO improves to within 1.5 dB from the lower bound (single-user MIMO). To verify the proposed codebook design, widely studied Gaussian random codebooks [11], [2] are considered for comparison. Note that though the overall discussion in this chapter has generally been on LTE and LTE-Advanced framework, the proposed feedback and precoding design can serve as a guideline for multi-user MIMO modes in any other modern wireless system which employs limited feedback schemes for CSIT acquisition.

2. LTE system model

2.1. LTE - A brief overview

In 3GPP LTE, a 2×2 configuration for MIMO is assumed as the baseline configuration, however configurations with four transmit or receive antennas are also foreseen and reflected in the specifications [17]. LTE restricts the transmission of maximum of two codewords in the downlink which can be mapped onto different layers where one codeword represents an output from the channel encoder. Number of layers available for the transmission is equal to the rank of the channel matrix (maximum 4). In this chapter, we restrict ourselves to the baseline configuration with the eNodeB (LTE notation for the base station) equipped with 2 antennas while we consider single and dual antenna user equipments (UEs). Physical layer technology employed for the downlink in LTE is OFDMA combined with bit interleaved coded modulation (BICM) [4]. Several different transmission bandwidths are possible, ranging from 1.08 MHz to 19.8 MHz with the constraint of being a multiple of 180 kHz. Resource Blocks (RBs) are defined as groups of 12 consecutive resource elements (REs - LTE notation for the subcarriers) with a bandwidth of 180 kHz thereby leading to the constant RE spacing of 15 kHz. Approximately 4 RBs form a subband and the feedback is generally done on subband basis. Seven operation modes are specified in the downlink of LTE, however, we shall focus on the following four modes:

- **Transmission mode 2.** Fall-back transmit diversity. Transmission rank is 1, i.e. one codeword is transmitted by the eNodeB. Employs Alamouti space-time or space-frequency codes [1].
- **Transmission mode 4.** Closed-loop spatial multiplexing. Transmission rank is 2, i.e. two codewords are transmitted by the eNodeB to the UE in the single-user MIMO mode. UEs need to have minimum of two antennas.
- **Transmission mode 5.** Multi-user MIMO mode. Supports only rank-1 transmission, i.e. one codeword for each UE.
- **Transmission mode 6.** Closed-loop precoding for rank-1 transmission, i.e. one codeword for the UE in the single-user MIMO mode.

In the case of transmit diversity and closed-loop precoding, one codeword (data stream) is transmitted to each UE using Alamouti code in the former case and LTE precoders in the latter case. Time-frequency resources are orthogonal to the different UEs in these modes thereby avoiding interference in the system. However, in the multi-user MIMO mode, parallel codewords are transmitted simultaneously, one for each UE, sharing the same time-frequency resources. Note that LTE restricts the transmission of one codeword to each UE in the multi-user MIMO mode.

For closed-loop transmission modes (mode 4, 5 and 6), precoding mechanisms are employed at the transmit side with the objective of maximizing throughput. The precoding is selected and applied by the eNodeB to the data transmission to a target UE based on the channel feedback received from that UE. This feedback includes a precoding matrix indicator (PMI), a channel rank indicator (RI) and a channel quality indicator (CQI). PMI is an index in the codebook for the preferred precoder to be used by the eNodeB. The granularity for the computation and signaling of the precoding index can range from a couple of RBs to the full bandwidth. For transmission mode 5, the eNodeB selects the precoding matrix to induce high orthogonality between the codewords so that the interference between UEs is minimized. In transmission modes 4 and 6, the eNodeB selects the precoding vector/matrix such that codewords are transmitted to the corresponding UEs with maximum throughput.

In order to avoid excessive downlink signaling, transmission mode for each UE is configured semi-statically via higher layer signaling, i.e. it is not allowed for a UE to be scheduled in one subframe in the multi-user MIMO mode and in the next subframe in the single-user MIMO mode. For transmission modes 4, 5 and 6, low-resolution precoders are employed which are based on the principle of EGT. For the case of eNodeB with two antennas, LTE

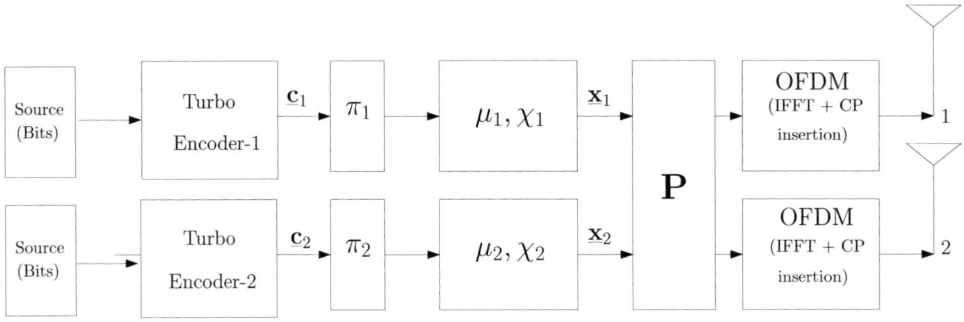

Figure 1. eNodeB in multi-user MIMO mode. π_1 denotes the random interleaver, μ_1 the labeling map and χ_1 the signal set for the codeword of UE-1. **P** indicates the precoding matrix.

proposes the use of following four precoders for transmission mode 5 and 6:

$$\mathbf{p} = \left\{ \frac{1}{\sqrt{4}} \begin{bmatrix} 1 \\ 1 \end{bmatrix}, \frac{1}{\sqrt{4}} \begin{bmatrix} 1 \\ -1 \end{bmatrix}, \frac{1}{\sqrt{4}} \begin{bmatrix} 1 \\ j \end{bmatrix}, \frac{1}{\sqrt{4}} \begin{bmatrix} 1 \\ -j \end{bmatrix} \right\} \tag{1}$$

The number of precoders increases to sixteen in the case of four transmit antennas however in this chapter we restrict to the case of two transmit antennas. For transmission mode 4, LTE proposes the use of following two precoder matrices on subband basis.

$$\mathbf{P} = \left\{ \frac{1}{\sqrt{4}} \begin{bmatrix} 1 & 1 \\ 1 & -1 \end{bmatrix}, \frac{1}{\sqrt{4}} \begin{bmatrix} 1 & 1 \\ j & -j \end{bmatrix} \right\} \tag{2}$$

Note that there is a possibility of swapping the columns in **P** but the swap must occur over the entire band.

2.2. System model

We first consider the system model for transmission mode 5, i.e. the multi-user MIMO mode in which the eNodeB transmits one codeword each to two single-antenna UEs using the same time-frequency resources. Transmitter block diagram is shown in Fig. 1. During the transmission for UE-1, the code sequence $\underline{\mathbf{c}}_1$ is interleaved by π_1 and is then mapped onto the signal sequence $\underline{\mathbf{x}}_1$. x_1 is the symbol of $\underline{\mathbf{x}}_1$ over a signal set $\chi_1 \subseteq \mathcal{C}$ with a Gray labeling map where $|\chi_1| = M_1$ and x_2 is the symbol of $\underline{\mathbf{x}}_2$ over signal set χ_2 where $|\chi_2| = M_2$. The bit interleaver for UE-1 can be modeled as $\pi_1 : k^{'} \rightarrow (k, i)$ where $k^{'}$ denotes the original ordering of the coded bits $c_{k'}$, k denotes the RE of the symbol $x_{1,k}$ and i indicates the position of the bit $c_{k'}$ in the symbol $x_{1,k}$. Note that each RE corresponds to a symbol from a constellation map χ_1 for UE-1 and χ_2 for UE-2. Selection of the normal or extended cyclic prefix (CP) for each OFDM symbol converts the downlink frequency-selective channel into parallel flat fading channels. Cascading IFFT at the eNodeB and FFT at the UE with the cyclic prefix extension, the transmission at the k-th RE for UE-1 in transmission mode 5 can be expressed as

$$y_{1,k} = \mathbf{h}_{1,k}^{\dagger}\mathbf{p}_{1,k}x_{1,k} + \mathbf{h}_{1,k}^{\dagger}\mathbf{p}_{2,k}x_{2,k} + z_{1,k} \tag{3}$$

where $y_{1,k}$ is the received symbol at UE-1 and $z_{1,k}$ is zero mean circularly symmetric complex white Gaussian noise of variance N_0. $x_{1,k}$ is the complex symbol for UE-1 with the variance σ_1^2 and $x_{2,k}$ is the complex symbol for UE-2 with the variance σ_2^2. $\mathbf{h}_{n,k}^{\dagger} \in \mathbb{C}^{1\times 2}$ symbolizes the spatially uncorrelated flat Rayleigh fading MISO channel from eNodeB to the n-th UE ($n = 1, 2$) at the k-th RE. Its elements can therefore be modeled as independent and identically distributed (iid) zero mean circularly symmetric complex Gaussian random variables with a variance of 0.5 per dimension. Note that $\mathbb{C}^{1\times 2}$ denotes a 2-dimensional complex space. $\mathbf{p}_{n,k}$ denotes the precoding vector for the n-th UE at the k-th RE and is given by (1). For the dual antenna UEs, the system equation for transmission mode 5 is modified as

$$\mathbf{y}_{1,k} = \mathbf{H}_{1,k}\left[\mathbf{p}_{1,k}x_{1,k} + \mathbf{p}_{2,k}x_{2,k}\right] + \mathbf{z}_{1,k} \tag{4}$$

where $\mathbf{y}_{1,k}, \mathbf{z}_{1,k} \in \mathbb{C}^{2\times 1}$ are the vectors of the received symbols and circularly symmetric complex white Gaussian noise of double-sided power spectral density $N_0/2$ at the 2 receive antennas of UE-1 respectively. $\mathbf{H}_{1,k} \in \mathbb{C}^{2\times 2}$ is the channel matrix from eNodeB to UE-1.

In transmission mode 6, only one UE will be served in one time-frequency resource. Therefore the system equation for single-antenna UEs at the k-th RE is given as

$$y_k = \mathbf{h}_k^{\dagger}\mathbf{p}_k x_k + z_k \tag{5}$$

where $\mathbf{p}_k$ is given by (1). For the dual antenna UEs, the system equation for mode 6 is modified as

$$\mathbf{y}_k = \mathbf{H}_k\mathbf{p}_k x_k + \mathbf{z}_k \tag{6}$$

3. Multi-user MIMO mode

We now look at the effectiveness of the low-resolution LTE precoders for the multi-user MIMO mode. We first consider a geometric scheduling strategy [8] based on the selection of UEs with orthogonal precoders.

3.1. Scheduling strategy

As the processing at the UE is performed on a RE basis for each received OFDM symbol, the dependency on RE index can be ignored for notational convenience. The system equation for the case of single-antenna UEs for the multi-user mode is

$$y_1 = \mathbf{h}_1^{\dagger}\mathbf{p}_1 x_1 + \mathbf{h}_1^{\dagger}\mathbf{p}_2 x_2 + z_1 \tag{7}$$

The scheduling strategy is based on the principle of maximizing the desired signal strength while minimizing the interference strength. As the decision to schedule a UE in the single-user MIMO, multi-user MIMO or transmit diversity mode will be made by the eNodeB, each UE would feedback the precoder which maximizes its received signal strength. So this selected precoder by the UE would be the one closest to its matched filter (MF) precoder in terms of the Euclidean distance.

For the multi-user MIMO mode, the eNodeB needs to ensure good channel separation between the co-scheduled UEs. Therefore the eNodeB schedules two UEs on the same RBs which have requested opposite (orthogonal) precoders, i.e. the eNodeB selects as the second UE to be served in each group of allocatable RBs, one of the UEs whose requested precoder $\mathbf{p}_2$ is 180° out of phase from the precoder $\mathbf{p}_1$ of the first UE to be served on the same RBs. So if UE-1 has requested $\mathbf{p}_1 = \frac{1}{\sqrt{4}}\begin{bmatrix} 1 \\ q \end{bmatrix}, \quad q \in \{\pm 1, \pm j\}$, then eNodeB selects the second UE which has requested $\mathbf{p}_2 = \frac{1}{\sqrt{4}}\begin{bmatrix} 1 \\ -q \end{bmatrix}$. This transmission strategy also remains valid also for the case of dual-antenna UEs where the UEs feedback the indices of the precoding vectors which maximize the strength of their desired signals, i.e. $\|\mathbf{H}\mathbf{p}\|^2$. For the multi-user MIMO mode, the eNodeB schedules two UEs on the same RE which have requested 180° out of phase precoders. The details of this geometric scheduling strategy can be found in [7].

Though this precoding and scheduling strategy would ensure minimization of the interference under the constraint of low-resolution LTE precoders, the residual interference would still be significant. Single-user detection i.e. Gaussian assumption of the residual interference and its subsequent absorption in noise would lead to significant degradation in the performance. On the other hand, this residual interference is actually discrete belonging to a finite alphabet and its structure can be exploited in the detection process. However intelligent detection based on its exploitation comes at the cost of enhanced complexity. Here we propose a low-complexity interference-aware receiver structure [9] which on one hand reduces one complex dimension of the system while on the other hand, it exploits the interference structure in the detection process.

3.2. Low-complexity interference-aware receiver

First we consider the case of single-antenna UEs. Soft decision of the bit $c_{k'}$ of x_1, also known as log-likelihood ratio (LLR), is given as

$$\mathsf{LLR}_1^i\left(c_{k'}|y_1,\mathbf{h}_1^\dagger,\mathbf{P}\right) = \log \frac{p\left(c_{k'}=1|y_1,\mathbf{h}_1^\dagger,\mathbf{P}\right)}{p\left(c_{k'}=0|y_1,\mathbf{h}_1^\dagger,\mathbf{P}\right)} \tag{8}$$

We introduce the notation $\Lambda_1^i\left(y_1,c_{k'}\right)$ for the bit metric which is developed on the lines similar to the equations (7) and (9) in [4], i.e.

$$\begin{aligned}
\Lambda_1^i\left(y_1, c_{k'}\right) &= \log p\left(c_{k'}|y_1, \mathbf{h}_1^{\dagger}, \mathbf{P}\right) \\
&\approx \log p\left(y_1|c_{k'}, \mathbf{h}_1^{\dagger}, \mathbf{P}\right) \\
&= \log \sum_{x_1 \in \chi_{1,c_{k'}}^i} \sum_{x_2 \in \chi_2} p\left(y_1|x_1, x_2, \mathbf{h}_1^{\dagger}, \mathbf{P}\right) \\
&\approx \min_{x_1 \in \chi_{1,c_{k'}}^i, x_2 \in \chi_2} \frac{1}{N_0}\left|y_1 - \mathbf{h}_1^{\dagger}\mathbf{p}_1 x_1 - \mathbf{h}_1^{\dagger}\mathbf{p}_2 x_2\right|^2
\end{aligned} \tag{9}$$

where $\chi_{1,c_{k'}}^i$ denotes the subset of the signal set $x_1 \in \chi_1$ whose labels have the value $c_{k'} \in \{0,1\}$ in the position i. Here we have used the log-sum approximation, i.e. $\log \sum_j z_j = \max_j \log z_j$ and this bit metric is therefore termed as max log MAP bit metric. As LLR is the difference of two bit metrics and these will be decoded using a conventional soft-decision Viterbi algorithm, $\frac{1}{N_0}$ (a common scaling factor to all LLRs) can be ignored thereby leading to

$$\begin{aligned}
\Lambda_1^i\left(y_1, c_{k'}\right) &\approx \min_{x_1 \in \chi_{1,c_{k'}}^i, x_2 \in \chi_2} \left|y_1 - \mathbf{h}_1^{\dagger}\mathbf{p}_1 x_1 - \mathbf{h}_1^{\dagger}\mathbf{p}_2 x_2\right|^2 \\
&= \min_{x_1 \in \chi_{1,c_{k'}}^i, x_2 \in \chi_2} \left\{|y_1|^2 + \left|\mathbf{h}_1^{\dagger}\mathbf{p}_1 x_1\right|^2 + \left|\mathbf{h}_1^{\dagger}\mathbf{p}_2 x_2\right|^2 - 2\left(\mathbf{h}_1^{\dagger}\mathbf{p}_1 x_1 y_1^*\right)_R + 2\left(\rho_{12} x_1^* x_2\right)_R - 2\left(\mathbf{h}_1^{\dagger}\mathbf{p}_2 x_2 y_1^*\right)_R\right\}
\end{aligned} \tag{10}$$

where $\rho_{12} = \left(\mathbf{h}_1^{\dagger}\mathbf{p}_1\right)^* \mathbf{h}_1^{\dagger}\mathbf{p}_2$ indicates the cross correlation between the two effective channels. Here we have used the relation $|a-b|^2 = |a|^2 + |b|^2 - 2\left(a^* b\right)_R$ where the subscript $(.)_R$ indicates the real part. Note that the complexity of the calculation of bit metric (10) is $\mathcal{O}\left(|\chi_1|\,|\chi_2|\right)$.

In (10), we now introduce two terms as the outputs of MF, i.e. $\bar{y}_1 = \left(\mathbf{h}_1^{\dagger}\mathbf{p}_1\right)^* y_1$ and $\bar{y}_2 = \left(\mathbf{h}_1^{\dagger}\mathbf{p}_2\right)^* y_1$. Ignoring $|y_1|^2$ (independent of the minimization operation), the bit metric is written as

$$\Lambda_1^i\left(y_1, c_{k'}\right) \approx \min_{x_1 \in \chi_{1,c_{k'}}^i, x_2 \in \chi_2} \left\{\left|\mathbf{h}_1^{\dagger}\mathbf{p}_1 x_1\right|^2 + \left|\mathbf{h}_1^{\dagger}\mathbf{p}_2 x_2\right|^2 - 2\left(\bar{y}_1^* x_1\right)_R + 2\psi_A x_{2,R} + 2\psi_B x_{2,I}\right\} \tag{11}$$

where

$$\begin{aligned}
\psi_A &= \rho_{12,R} x_{1,R} + \rho_{12,I} x_{1,I} - \bar{y}_{2,R} \\
\psi_B &= \rho_{12,R} x_{1,I} - \rho_{12,I} x_{1,R} - \bar{y}_{2,I}
\end{aligned}$$

Note that the subscript $(.)_I$ indicates the imaginary part.

For x_1 and x_2 belonging to equal energy alphabets, $\left|\mathbf{h}_1^\dagger \mathbf{p}_1 x_1\right|^2$ and $\left|\mathbf{h}_1^\dagger \mathbf{p}_2 x_2\right|^2$ can be ignored as they are independent of the minimization operation. The values of $x_{2,R}$ and $x_{2,I}$ which minimize (11) need to be in the opposite directions of ψ_A and ψ_B respectively thereby avoiding search on the alphabets of x_2 and reducing one complex dimension in the detection, i.e.

$$\Lambda_1^i\left(y_1, c_{k'}\right) \approx \min_{x_1 \in \mathcal{X}^i_{1,c_{k'}}} \left\{-2\overline{y}_{1,R} x_{1,R} - 2\overline{y}_{1,I} x_{1,I} - 2\left|\psi_A\right|\left|x_{2,R}\right| - 2\left|\psi_B\right|\left|x_{2,I}\right|\right\} \tag{12}$$

As an example we consider the case of QPSK for which the values of $x_{2,R}$ and $x_{2,I}$ are $\left[\pm\frac{\sigma_2}{\sqrt{2}}\right]$, so the bit metric is written as

$$\Lambda_1^i\left(y_1, c_{k'}\right) \approx \min_{x_1 \in \mathcal{X}^i_{1,c_{k'}}} \left\{-2\overline{y}_{1,R} x_{1,R} - 2\overline{y}_{1,I} x_{1,I} - \sqrt{2}\sigma_2\left|\psi_A\right| - \sqrt{2}\sigma_2\left|\psi_B\right|\right\} \tag{13}$$

For x_1 and x_2 belonging to non-equal energy alphabets, the bit metric is same as (13) but $\left|\mathbf{h}_1^\dagger \mathbf{p}_1 x_1\right|^2$ and $\left|\mathbf{h}_1^\dagger \mathbf{p}_2 x_2\right|^2$ can no longer be ignored thereby leading to

$$\begin{aligned}\Lambda_1^i\left(y_1, c_{k'}\right) \approx \min_{x_1 \in \mathcal{X}^i_{1,c_{k'}}} \Bigg\{&\left|\mathbf{h}_1^\dagger \mathbf{p}_1\right|^2 \left|x_{1,R}\right|^2 + \left|\mathbf{h}_1^\dagger \mathbf{p}_1\right|^2 \left|x_{1,I}\right|^2 + \left|\mathbf{h}_1^\dagger \mathbf{p}_2\right|^2 \left|x_{2,R}\right|^2 + \left|\mathbf{h}_1^\dagger \mathbf{p}_2\right|^2 \left|x_{2,I}\right|^2 - \\ &2\overline{y}_{1,R} x_{1,R} - 2\overline{y}_{1,I} x_{1,I} - 2\left|\psi_A\right|\left|x_{2,R}\right| - 2\left|\psi_B\right|\left|x_{2,I}\right|\Bigg\}\end{aligned} \tag{14}$$

Note that the minimization is independent of χ_2 though x_2 appears in the bit metric. The reason of this independence is as follows. The decision regarding the signs of $x_{2,R}$ and $x_{2,I}$ in (14) will be taken in the same manner as for the case of equal energy alphabets. For finding their magnitudes that minimize the bit metric (14), it is the minimization problem of a quadratic function, i.e. differentiating (14) w.r.t $\left|x_{2,R}\right|$ and $\left|x_{2,I}\right|$ to find the global minimas which are given as

$$\left|x_{2,R}\right| \rightarrow \frac{\left|\psi_A\right|}{\left|\mathbf{h}_1^\dagger \mathbf{p}_2\right|^2}, \qquad \left|x_{2,I}\right| \rightarrow \frac{\left|\psi_B\right|}{\left|\mathbf{h}_1^\dagger \mathbf{p}_2\right|^2} \tag{15}$$

where $\rightarrow$ indicates the discretization process in which amongst the finite available points of $x_{2,R}$ and $x_{2,I}$, the point closest to the calculated continuous value is selected. So if x_2 belongs to QAM256, then instead of searching 256 constellation points for the minimization of (14), the metric (15) reduces it to merely two operations thereby trimming down one complex dimension in the detection, i.e. the detection complexity is independent of $|\chi_2|$ and reduces to $\mathcal{O}(|\chi_1|)$.

As a particular example of the discretization of continuous values in (15), we consider the case of x_2 belonging to QAM16 . The values of $x_{2,R}$ and $x_{2,I}$ for the case of QAM16 are $\left[\pm\frac{\sigma_2}{\sqrt{10}}, \pm\frac{3\sigma_2}{\sqrt{10}}\right]$ so their magnitudes in (14) are given as

$$|x_{2,R}| = \sigma_2 \frac{1}{\sqrt{10}} \left(2 + (-1)^{I\left(|\psi_A| < \sigma_2 \frac{2\left|\mathbf{h}_1^\dagger \mathbf{p}_2\right|^2}{\sqrt{10}}\right)}\right)$$

$$|x_{2,I}| = \sigma_2 \frac{1}{\sqrt{10}} \left(2 + (-1)^{I\left(|\psi_B| < \sigma_2 \frac{2\left|\mathbf{h}_1^\dagger \mathbf{p}_2\right|^2}{\sqrt{10}}\right)}\right) \tag{16}$$

and $I(.)$ is the indicator function defined as

$$I(a < b) = \begin{cases} 1 \text{ if } a < b \\ 0 \text{ otherwise} \end{cases}$$

Now we look at the receiver structure for the case of dual-antenna UEs. The system equation for UE-1 (ignoring the RE index) is

$$\mathbf{y}_1 = \mathbf{H}_1 \left[\mathbf{p}_1 x_1 + \mathbf{p}_2 x_2\right] + \mathbf{z}_1 \tag{17}$$

The receiver structure would remain same with $\mathbf{h}_1^\dagger$ being replaced by $\mathbf{H}_1$, i.e. the channel from eNodeB to the two antennas of UE-1. Subsequently $\bar{y}_1 = (\mathbf{H}_1\mathbf{p}_1)^\dagger \mathbf{y}_1$ and $\bar{y}_2 = (\mathbf{H}_1\mathbf{p}_2)^\dagger \mathbf{y}_1$ are the MF outputs while $\rho_{12} = (\mathbf{H}_1\mathbf{p}_1)^\dagger \mathbf{H}_1\mathbf{p}_2$ is the cross-correlation between two effective channels.

For comparison purposes, we also consider the case of single-user receiver, for which the bit metric is given as

$$\Lambda_1^i\left(y_1, c_{k'}\right) \approx \min_{x_1 \in \chi^i_{1,c_{k'}}} \left\{ \frac{1}{\left(|\rho_{12}|^2 \sigma_2^2 + \left|\mathbf{h}_1^\dagger \mathbf{p}_1\right|^2 N_0\right)} \left|\bar{y}_1 - \left|\mathbf{h}_1^\dagger \mathbf{p}_1\right|^2 x_1\right|^2 \right\} \tag{18}$$

Table 1 compares the complexities of different receivers in terms of the number of real-valued multiplications and additions for getting all LLR values per RE/subcarrier. Note that n_r denotes the number of receive antennas. This complexity analysis is independent of the number of transmit antennas as the operation of finding effective channels bears same

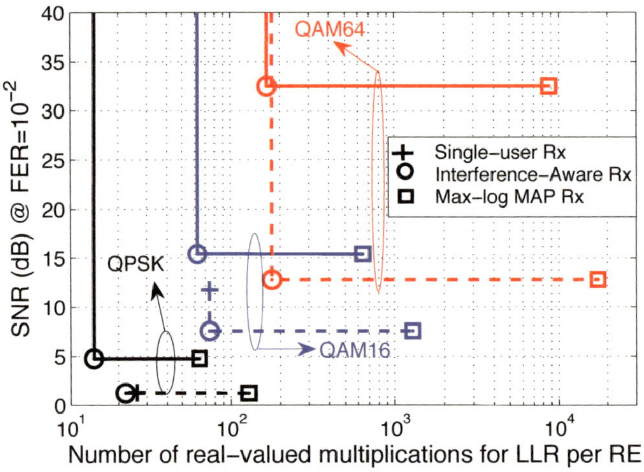

Figure 2. eNodeB has two antennas. Continuous lines indicate the case of single-antenna UEs while dashed lines indicate dual-antenna UEs. 3GPP LTE rate 1/2 punctured turbo code is used. Simulation settings are same as in the first part of Sec.7.

Receiver	Real Multiplications	Real Additions
Interference-aware receiver (Equal energy alphabets)	$8n_r + 2\sqrt{M} + 2M$	$8n_r + 10M + \log(M) - 4$
Interference-aware receiver (Non Equal energy alphabets)	$12n_r + 4M + \frac{7}{2}\sqrt{M}$	$12n_r + 18M + \log(M) - 6$
Max-log MAP receiver	$2M^2n_r + 8Mn_r$	$6M^2n_r + 4Mn_r + \log(M) - M^2$
Single-user receiver (Equal energy alphabets)	$10n_r + 6$	$10n_r - 3$
Single-user receiver (Non Equal energy alphabets)	$10n_r + 3M + \sqrt{M}/2 + 4$	$10n_r + 3M + \log(M) - 3$

Table 1. Comparison of receivers complexity

complexity in all receiver structures. Moreover UEs can also directly estimate their effective channels if the pilot signals are also precoded. The comparison shows that the complexity of the interference-aware receiver is of the same order as of single-user receiver while it is far less than the complexity of the max log MAP receiver. Fig. 2 further shows the performance-complexity trade off of different receivers for multi-user MIMO mode in LTE. The performance of the receivers is measured in terms of the SNR at the frame error rate (FER) of 10^{-2} whereas the complexity is determined from Table.1. It shows that the performance of the single-user receiver is severely degraded as compared to that of the interference-aware receiver. In most cases, the single-user receiver fails to achieve the requisite FER in the considered SNR range. On the other hand, interference-aware receiver achieves same performance as max log MAP receiver but with much reduced complexity.

The interference-aware receiver is therefore not only characterized by low complexity but it also resorts to intelligent detection by exploiting the structure of residual interference. Moreover, this receiver structure being based on the MF outputs and devoid of any division operation can be easily implemented in the existing hardware. However the proposed receiver needs both the channel knowledge and the constellation of interference (co-scheduled UE). As the UE already knows its own channel from the eNodeB and the requested precoder, it can determine the effective channel of the interference based on the geometric scheduling algorithm, i.e. the precoder of the co-scheduled UE is 180° out of phase of its own precoder. Consequently there is no additional complexity in utilizing this receiver

http://dx.doi.org/10.5772/57134

structure as compared to using single-user receivers except that the UE needs to know the constellation of interference.

4. Information theoretic perspective

Sum rate of the downlink channel is given as

$$\mathcal{I} = I\left(Y_1; X_1 | \mathbf{h}_1^\dagger, \mathbf{P}\right) + I\left(Y_2; X_2 | \mathbf{h}_2^\dagger, \mathbf{P}\right) \tag{19}$$

where $\mathbf{P} = [\mathbf{p}_1\ \mathbf{p}_2]$ is the precoder matrix, $I\left(Y_1; X_1 | \mathbf{h}_1^\dagger, \mathbf{P}\right)$ is the mutual information of UE-1 once it sees interference from UE-2 and $I\left(Y_2; X_2 | \mathbf{h}_2^\dagger, \mathbf{P}\right)$ is the mutual information of UE-2 once it sees interference from UE-1. Note that Y_1 is the received symbol at UE-1 while X_1 is the symbol transmitted by the eNodeB to UE-1. Note that interference is present in the statistics of Y_1 and Y_2. No sophisticated power allocation is employed to the two streams as the downlink control information (DCI) in the multi-user mode in LTE includes only 1-bit power offset information, indicating whether a 3 dB transmit power reduction should be assumed or not. We therefore consider equal-power distribution between the two streams. For the calculation of mutual information, we deviate from the unrealistic Gaussian assumption for the alphabets and consider them from discrete constellations. The derivations of the mutual information expressions for the case of finite alphabets have been relegated to Appendix-A for simplicity and lucidity.

We focus on the LTE precoders but to analyze the degradation caused by the low-level quantization and the characteristic of EGT of these precoders, we also consider some other transmission strategies. Firstly we consider unquantized MF precoder [27] which is given as

$$\mathbf{p} = \frac{1}{\sqrt{|h_{11}|^2 + |h_{21}|^2}} \begin{bmatrix} h_{11} \\ h_{21} \end{bmatrix} \tag{20}$$

For EGT, the unquantized MF precoder is given as

$$\mathbf{p} = \frac{1}{\sqrt{2}} \begin{bmatrix} 1 \\ h_{11}^* h_{21} / |h_{11}| \, |h_{21}| \end{bmatrix} \tag{21}$$

To be fair in comparison with the geometric scheduling algorithm for multi-user MIMO in LTE, we introduce a geometric scheduling algorithm for unquantized precoders. We divide the spatial space into 4 quadrants according to the spatial angle between $\mathbf{h}_1^\dagger$ and $\mathbf{h}_2^\dagger$ which is given as

$$\phi = \cos^{-1}\left(\frac{\left|\mathbf{h}_1^\dagger \mathbf{h}_2\right|}{\|\mathbf{h}_1\| \, \|\mathbf{h}_2\|}\right) \qquad 0^\circ \leq \phi \leq 90^\circ \tag{22}$$

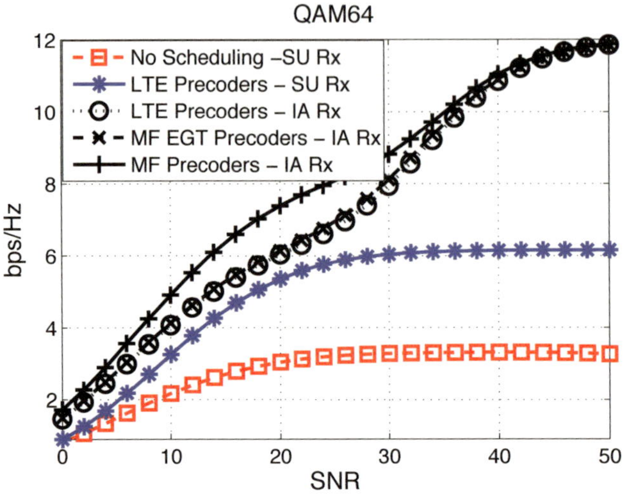

Figure 3. Sum rates of different transmission schemes for the downlink channel with dual-antenna eNodeB and 2 single-antenna UEs. 'No Scheduling - SU Rx' indicates the case once the eNodeB uses the LTE precoders without employing the geometric scheduling strategy. In all other cases, the eNodeB employs the geometric scheduling strategy along with the LTE precoders, MF EGT precoders and MF precoders. SU Rx indicates the cases when UEs employ single-user detection while IA Rx indicates the cases when UEs resort to the intelligent detection by employing the low-complexity interference-aware receivers.

The geometric scheduling algorithm ensures that the eNodeB chooses the second UE to be served on the same RE as the first UE such that their channels $\mathbf{h}_1^\dagger$ and $\mathbf{h}_2^\dagger$ lie in the opposite quadrants.

Fig. 3 shows the sum rates of a broadcast channel with the dual-antenna eNodeB and 2 single-antenna UEs for QAM64 alphabets. SNR is the transmit SNR, i.e. $\frac{\sigma_1^2\|\mathbf{p}_1\|^2+\sigma_2^2\|\mathbf{p}_2\|^2}{N_0}$ whereas the two UEs have equal power distribution, i.e. $\sigma_1^2 = \sigma_2^2$. MF and MF EGT precoders are the unquantized precoders given in (20) and (21) respectively while LTE precoders are the quantized precoders given in (1). The sum rates of unquantized precoders along with those of LTE quantized precoders are shown for the case of single-user receivers and for the case of low-complexity interference-aware receivers. The results show that under the proposed transmission strategy, the sum rate can be significantly improved (unbounded in SNR) if the low-complexity interference-aware receivers are used as compared to the case when the UEs resort to sub-optimal single-user detection where rates are bounded (in SNR). The behavior of single-user detection is attributed to the fact that this detection strategy considers interference as noise so the SINR is low once no geometric scheduling has been employed by the eNodeB while the SINR improves due to the reduction of interference once geometric scheduling is employed. However the rates remain bounded in the SNR if the UEs resort to the single-user detection which is due to the fact that increasing the SNR (transmit SNR) also increases the interference strength thereby bounding the SINR at high values of the transmit SNR. On the other hand, there is significant improvement in the sum rate once UEs resort to intelligent detection by employing the low-complexity interference-aware receivers. In this case, the sum rate is unbounded if the rate (constellation size) of each UE is adapted with the SNR. Note that the quantized CSIT (LTE precoders) appears to have no effect at high SNR once UEs resort to intelligent interference-aware detection. This behavior is because the rate is not adapted with the SNR in these simulations, i.e. the constellation size is fixed to QAM64 and is not increased with the increase in the SNR. At high SNR, the rate of each UE gets saturated to its constellation size (6 bits for QAM64) if the UE resorts to intelligent interference-aware detection. However the approach to this saturation point (slope of the rate curve) depends on the quantization of channel information.

Another interesting result is the effect of the two characteristics of LTE precoders, i.e. low-resolution and EGT. There is a slight improvement in the sum rate at medium SNR when the restriction of low-resolution (LTE quantized precoders) is relaxed, i.e. eNodeB employs MF EGT precoders however there is a significant improvement in the sum rate when the restriction of EGT is eliminated, i.e the eNodeB employs MF precoders. This shows that the loss in spectral efficiency due to the employment of LTE precoders is mainly attributed to the EGT rather than their low resolution (quantization).

5. Performance analysis

We now focus on the EGT characteristic of the LTE precoders and carry out the performance analysis of the EGT in single-user and multi-user MIMO systems. We restrict to the case of single-antenna UEs while the eNodeB has two antennas.

5.1. Single-user MIMO

For the single-user case, the received signal at the k-th RE is given by

$$y_{1,k} = \mathbf{h}_{1,k}^{\dagger}\mathbf{p}_{1,k}x_{1,k} + z_{1,k} \tag{23}$$

For EGT, the precoder vector is given by $\mathbf{p}_{1,k} = \frac{1}{\sqrt{2}}\begin{bmatrix}1 & \frac{h_{21,k}h_{11,k}^{*}}{|h_{21,k}||h_{11,k}|}\end{bmatrix}^{T}$. So the received signal after normalization by $\frac{h_{11,k}}{|h_{11,k}|}$ is given by

$$y_{1,k}^{N} = \frac{1}{\sqrt{2}}\left(|h_{11,k}| + |h_{21,k}|\right)x_{1,k} + \frac{h_{11,k}}{|h_{11,k}|}z_{1,k} \tag{24}$$

where $y_{1,k}^{N} = \frac{h_{11,k}}{|h_{11,k}|}y_{1,k}$. The max log MAP bit metric [4] for the bit $c_{k'}$ can be written as

$$\Lambda_1^i\left(y_k, c_{k'}\right) \approx \min_{x_1 \in \chi_{1,c_{k'}}^i}\left[\frac{1}{N_0}\left|y_{1,k}^{N} - \frac{1}{\sqrt{2}}\left(|h_{11,k}| + |h_{21,k}|\right)x_1\right|^2\right] \tag{25}$$

The conditional PEP i.e $P\left(\underline{\mathbf{c}}_1 \rightarrow \hat{\underline{\mathbf{c}}}_1|\mathbf{h}_1\right)$ is given as

$$P\left(\underline{\mathbf{c}}_1 \rightarrow \hat{\underline{\mathbf{c}}}_1|\overline{\mathbf{H}}_1\right) = P\left(\sum_{k'} \min_{x_1 \in \mathcal{X}^i_{1,c_{k'}}} \frac{1}{N_0}\left|y^N_{1,k} - \frac{1}{\sqrt{2}}\left(\left|h_{11,k}\right| + \left|h_{21,k}\right|\right) x_1\right|^2 \right.$$
$$\left. \geq \sum_{k'} \min_{x_1 \in \mathcal{X}^i_{1,\hat{c}_{k'}}} \frac{1}{N_0}\left|y^N_{1,k} - \frac{1}{\sqrt{2}}\left(\left|h_{11,k}\right| + \left|h_{21,k}\right|\right) x_1\right|^2 \middle| \overline{\mathbf{H}}_1\right) \tag{26}$$

where $\overline{\mathbf{H}}_1$ indicates the complete channel from the eNodeB to UE-1 for the transmission of the codeword $\underline{\mathbf{c}}_1$. Assume $d\left(\underline{\mathbf{c}}_1 - \hat{\underline{\mathbf{c}}}_1\right) = d_{free}$ for $\underline{\mathbf{c}}_1$ and $\hat{\underline{\mathbf{c}}}_1$ under consideration for the PEP analysis, which is the worst case scenario between any two codewords. Therefore, the inequality on the right hand side of (26) shares the same terms on all but d_{free} summation points and the summations can be simplified to only d_{free} terms for which $\hat{c}_{k'} = \bar{c}_{k'}$. Let's denote

$$\tilde{x}_{1,k} = \arg \min_{x_1 \in \mathcal{X}^i_{1,c_{k'}}} \frac{1}{N_0}\left|y^N_{1,k} - \frac{1}{\sqrt{2}}\left(\left|h_{11,k}\right| + \left|h_{21,k}\right|\right) x_1\right|^2$$
$$\hat{x}_{1,k} = \arg \min_{x_1 \in \mathcal{X}^i_{1,\bar{c}_{k'}}} \frac{1}{N_0}\left|y^N_{1,k} - \frac{1}{\sqrt{2}}\left(\left|h_{11,k}\right| + \left|h_{21,k}\right|\right) x_1\right|^2 \tag{27}$$

As $\frac{1}{N_0}\left|y^N_{1,k} - \frac{1}{\sqrt{2}}\left(\left|h_{11,k}\right| + \left|h_{21,k}\right|\right) x_{1,k}\right|^2 \geq \frac{1}{N_0}\left|y^N_{1,k} - \frac{1}{\sqrt{2}}\left(\left|h_{11,k}\right| + \left|h_{21,k}\right|\right) \tilde{x}_{1,k}\right|^2$, this leads to PEP being given as

$$P\left(\underline{\mathbf{c}}_1 \rightarrow \hat{\underline{\mathbf{c}}}_1|\overline{\mathbf{H}}_1\right) \leq P\left(\sum_{k,d_{free}} \frac{1}{N_0}\left|y^N_{1,k} - \frac{1}{\sqrt{2}}\left(\left|h_{11,k}\right| + \left|h_{21,k}\right|\right) x_{1,k}\right|^2 \right.$$
$$\left. \geq \sum_{k,d_{free}} \frac{1}{N_0}\left|y^N_{1,k} - \frac{1}{\sqrt{2}}\left(\left|h_{11,k}\right| + \left|h_{21,k}\right|\right) \hat{x}_{1,k}\right|^2 \middle| \overline{\mathbf{H}}_1\right) \tag{28}$$

$$= P\left(\sum_{k,d_{free}} \frac{\sqrt{2}\left(\left|h_{11,k}\right| + \left|h_{21,k}\right|\right)}{N_0}\left(z^*_{1,k}\left(\hat{x}_{1,k} - x_{1,k}\right)\right)_R \right.$$
$$\left. \geq \sum_{k,d_{free}} \frac{1}{2N_0}\left(\left|h_{11,k}\right| + \left|h_{21,k}\right|\right)^2 \left|\hat{x}_{1,k} - x_{1,k}\right|^2\right)$$
$$= Q\left(\sqrt{\sum_{k,d_{free}} \frac{1}{4N_0}\left(\left|h_{11,k}\right| + \left|h_{21,k}\right|\right)^2 \left|\left(x_{1,k} - \hat{x}_{1,k}\right)\right|^2}\right)$$
$$\leq \frac{1}{2}\exp\left(-\sum_{k,d_{free}} \frac{1}{8N_0}\left(\left|h_{11,k}\right| + \left|h_{21,k}\right|\right)^2 d^2_{1,\min}\right)$$
$$= \frac{1}{2}\prod_{k,d_{free}} \exp\left(-\frac{1}{8N_0}\left(\left|h_{11,k}\right| + \left|h_{21,k}\right|\right)^2 d^2_{1,\min}\right) \tag{29}$$

where we have used Chernoff bound $Q(x) \leq \frac{1}{2}\exp\left(\frac{-x^2}{2}\right)$. Averaging over channel leads to

$$\begin{aligned} P(\underline{\mathbf{c}}_1 \rightarrow \hat{\underline{\mathbf{c}}}_1) &\leq \frac{1}{2} E_{\overline{\mathbf{H}}_1} \prod_{k,d_{free}} \exp\left(-\frac{1}{8N_0}\left(|h_{11,k}| + |h_{21,k}|\right)^2 d_{1,\min}^2\right) \\ &= \frac{1}{2} \prod_{k,d_{free}} E_{\mathbf{h}_{1,k}} \exp\left(\left(-\frac{\tilde{d}_{1,\min}^2}{4}\right) \frac{\left(|h_{11,k}| + |h_{21,k}|\right)^2 \sigma_1^2}{2N_0}\right) \end{aligned} \tag{30}$$

(30) follows from the channel independence at each RE which is the consequence of the interleaving operation. Here we have used the notation $d_{1,\min}^2 = \sigma_1^2 \tilde{d}_{1,\min}^2$ with $\tilde{d}_{1,\min}^2$ being the normalized minimum distance of the constellation χ_1. Using the moment generating function (MGF) of the SNR at the output of two branch EGC as per equations (2) and (23) in [21], PEP at high SNR is upper bounded as

$$\begin{aligned} P(\underline{\mathbf{c}}_1 \rightarrow \hat{\underline{\mathbf{c}}}_1) \leq \frac{1}{2} \prod_{d_{free}} \Bigg(& \frac{8\left(\frac{\sigma_1^2}{N_0}\right)^2 + \tilde{d}_{1,\min}^2 \left(\frac{\sigma_1^2}{N_0}\right)^3}{4\left(\frac{\sigma_1^2}{N_0}\right)^2 \left(2 + \frac{\sigma_1^2 \tilde{d}_{1,\min}^2}{4N_0}\right)^2} - \frac{\left(\frac{\tilde{d}_{1,\min}^2}{2\sqrt{2}}\right)\left(\frac{\sigma_1^2}{N_0}\right)}{\left(2 + \frac{\tilde{d}_{1,\min}^2}{2}\left(\frac{\sigma_1^2}{N_0}\right)\right)^{3/2}} \\ & \times \left[\pi - 2\sin^{-1}\left(\sqrt{\frac{\left(\frac{\sigma_1^2}{N_0}\right)^{-1} + \frac{\tilde{d}_{1,\min}^2}{4}}{2\left(\frac{\sigma_1^2}{N_0}\right)^{-1} + \frac{\tilde{d}_{1,\min}^2}{4}}}\right)\right] \\ & + \frac{4\left(\frac{\sigma_1^2}{N_0}\right)^2 \left(4 + \frac{\tilde{d}_{1,\min}^2}{2}\left(\frac{\sigma_1^2}{N_0}\right)\right)}{4\left(\frac{\sigma_1^2}{N_0}\right)^2 \left(2 + \frac{\tilde{d}_{1,\min}^2}{4}\left(\frac{\sigma_1^2}{N_0}\right)\right)^2 \left(2 + \frac{\tilde{d}_{1,\min}^2}{2}\left(\frac{\sigma_1^2}{N_0}\right)\right)} \Bigg) \end{aligned} \tag{31}$$

Using the identity $\cos^{-1}(x) = \frac{\pi}{2} - \sin^{-1}(x)$, we have

$$\pi - 2\sin^{-1}\left(\sqrt{\frac{\left(\frac{\sigma_1^2}{N_0}\right)^{-1} + \frac{\tilde{d}_{1,\min}^2}{4}}{2\left(\frac{\sigma_1^2}{N_0}\right)^{-1} + \frac{\tilde{d}_{1,\min}^2}{4}}}\right) = 2\cos^{-1}\left(\sqrt{\frac{\left(\frac{\sigma_1^2}{N_0}\right)^{-1} + \frac{\tilde{d}_{1,\min}^2}{4}}{2\left(\frac{\sigma_1^2}{N_0}\right)^{-1} + \frac{\tilde{d}_{1,\min}^2}{4}}}\right) \tag{32}$$

Taylor series expansion [10] of $\cos^{-1}\left(\sqrt{x}\right)$ is given as

$$\cos^{-1}\left(\sqrt{x}\right) = \sqrt{2-2\sqrt{x}}\sum_{k=0}^{\infty}\frac{\left(1-\sqrt{x}\right)^{k}(1/2)_{k}}{2^{k}\left(k!+2kk!\right)} \qquad \text{for } \left|-1+\sqrt{x}\right| < 2$$

where $x!$ is the factorial of x while $(x)_n$ is the Pochhammer symbol, i.e. $(x)_n = x(x+1)\cdots(x+n-1)$. For x closer to 1, a case that shall be occurring at high SNR in (32), first term will be dominant, i.e.

$$\cos^{-1}\left(\sqrt{\frac{\left(\frac{\sigma_1^2}{N_0}\right)^{-1}+\frac{d_{1,\min}^2}{4}}{2\left(\frac{\sigma_1^2}{N_0}\right)^{-1}+\frac{d_{1,\min}^2}{4}}}\right) \approx \sqrt{2-2\sqrt{\frac{\left(\frac{\sigma_1^2}{N_0}\right)^{-1}+\frac{d_{1,\min}^2}{4}}{2\left(\frac{\sigma_1^2}{N_0}\right)^{-1}+\frac{d_{1,\min}^2}{4}}}} \tag{33}$$

Taylor series expansion of $\sqrt{x}$ at $x=1$ is

$$\sqrt{x} = 1+\frac{x-1}{2}-\frac{(x-1)^2}{8}+\frac{(x-1)^3}{16}-\cdots$$

In the expansion of $\sqrt{\frac{\left(\frac{\sigma_1^2}{N_0}\right)^{-1}+\frac{d_{1,\min}^2}{4}}{2\left(\frac{\sigma_1^2}{N_0}\right)^{-1}+\frac{d_{1,\min}^2}{4}}}$, first two terms will be dominant at high SNR thereby leading to

$$\begin{aligned}
\sqrt{2-\sqrt{\frac{\left(\frac{\sigma_1^2}{N_0}\right)^{-1}+\frac{d_{1,\min}^2}{4}}{2\left(\frac{\sigma_1^2}{N_0}\right)^{-1}+\frac{d_{1,\min}^2}{4}}}} &\approx \sqrt{2-2\left(1+\frac{\frac{\left(\frac{\sigma_1^2}{N_0}\right)^{-1}+\frac{d_{1,\min}^2}{4}}{2\left(\frac{\sigma_1^2}{N_0}\right)^{-1}+\frac{d_{1,\min}^2}{4}}-1}{2}\right)} \\
&= \sqrt{-\left(\frac{\left(\frac{\sigma_1^2}{N_0}\right)^{-1}+\frac{d_{1,\min}^2}{4}}{2\left(\frac{\sigma_1^2}{N_0}\right)^{-1}+\frac{d_{1,\min}^2}{4}}-1\right)} \\
&= \frac{1}{\sqrt{2+\frac{d_{1,\min}^2}{4}\left(\frac{\sigma_1^2}{N_0}\right)}}
\end{aligned} \tag{34}$$

So rewriting (31), we get

$$P\left(\underline{\mathbf{c}}_1 \rightarrow \underline{\hat{\mathbf{c}}}_1\right) \leq \frac{1}{2}\prod_{d_{free}}\left(\frac{2}{\left(2+\frac{\breve{d}^2_{1,\min}}{4}\left(\frac{\sigma_1^2}{N_0}\right)\right)^2}+\frac{\breve{d}^2_{1,\min}\left(\frac{\sigma_1^2}{N_0}\right)}{4\left(2+\frac{\breve{d}^2_{1,\min}}{4}\left(\frac{\sigma_1^2}{N_0}\right)\right)^2}\right.$$
$$-\frac{2\left(\frac{\breve{d}^2_{1,\min}}{2\sqrt{2}}\right)\left(\frac{\sigma_1^2}{N_0}\right)}{\left(2+\frac{\breve{d}^2_{1,\min}}{2}\left(\frac{\sigma_1^2}{N_0}\right)\right)^{3/2}\left(2+\frac{\breve{d}^2_{1,\min}}{4}\left(\frac{\sigma_1^2}{N_0}\right)\right)^{1/2}}$$
$$\left.+\frac{\left(4+\frac{\breve{d}^2_{1,\min}}{2}\left(\frac{\sigma_1^2}{N_0}\right)\right)}{\left(2+\frac{\breve{d}^2_{1,\min}}{4}\left(\frac{\sigma_1^2}{N_0}\right)\right)^2\left(2+\frac{\breve{d}^2_{1,\min}}{2}\left(\frac{\sigma_1^2}{N_0}\right)\right)}\right) \tag{35}$$

At high SNR, second term converges to $\frac{4}{\breve{d}^2_{1,\min}\left(\frac{\sigma_1^2}{N_0}\right)}$ while the third term converges to $\frac{-4}{\breve{d}^2_{1,\min}\left(\frac{\sigma_1^2}{N_0}\right)}$. So PEP at high SNR is upper bounded as

$$\begin{aligned}P\left(\underline{\mathbf{c}}_1 \rightarrow \underline{\hat{\mathbf{c}}}_1\right) &\leq \frac{1}{2}\prod_{d_{free}}\left(\frac{32}{\left(\breve{d}^2_{1,\min}\left(\frac{\sigma_1^2}{N_0}\right)\right)^2}+\frac{16}{\left(\breve{d}^2_{1,\min}\left(\frac{\sigma_1^2}{N_0}\right)\right)^2}\right)\\ &= \frac{1}{2}\prod_{d_{free}}\left(\frac{48}{\left(\breve{d}^2_{1,\min}\left(\frac{\sigma_1^2}{N_0}\right)\right)^2}\right)\end{aligned} \tag{36}$$

where $\breve{d}^2_{1,\min}$ is the normalized minimum distance of the constellation χ_1, d_{free} is the free distance (minimum Hamming distance) of the code. Note that $\underline{\mathbf{c}}_1$ and $\underline{\hat{\mathbf{c}}}_1$ are the correct and error codewords respectively. (36) clearly shows full diversity of the EGT for single-user MIMO.

We now consider another approach to analyze the diversity order of EGT in the single-user MIMO transmission mode. We now focus on equal energy alphabets and consider the case of slow fading channel, i.e. the channel remains constant for the duration of one codeword. Rewriting the metric (28) in vector formulation i.e.

$$\sum_{k,d_{free}}\left|y^N_{1,k}-\frac{1}{\sqrt{2}}\left(|h_{11,k}|+|h_{21,k}|\right)x_{1,k}\right|^2=\left\|\mathbf{y}^N_1-\frac{1}{\sqrt{2}}\left(|h_{11}|+|h_{21}|\right)\mathbf{x}_1\right\|^2$$

So conditional PEP is given as

$$\begin{aligned}P\left(\underline{\mathbf{c}}_1 \to \hat{\underline{\mathbf{c}}}_1|\overline{\mathbf{H}}_1\right) &\leq P\left(\left\|\mathbf{y}_1^N - \frac{1}{\sqrt{2}}\left(|h_{11}|+|h_{21}|\right)\mathbf{x}_1\right\|^2 \geq \left\|\mathbf{y}_1^N - \frac{1}{\sqrt{2}}\left(|h_{11}|+|h_{21}|\right)\hat{\mathbf{x}}_1\right\|^2\right)\\ &= P\left(\Re\left(\left(\frac{1}{\sqrt{2}}\left(|h_{11}|+|h_{21}|\right)\mathbf{x}+\mathbf{z}_1\right)^{\dagger}\left(\mathbf{x}_1-\hat{\mathbf{x}}_1\right)\right)\leq 0\right)\\ &= P\left(\frac{1}{\sqrt{2}}\left(|h_{11}|+|h_{21}|\right)\left(\|\mathbf{x}_1\|^2 - \Re\left(\mathbf{x}_1^{\dagger}\hat{\mathbf{x}}_1\right)\right)+\Re\left(\mathbf{z}_1^{\dagger}\left(\mathbf{x}_1-\hat{\mathbf{x}}_1\right)\right)\leq 0\right)\end{aligned} \tag{37}$$

Using $\|\mathbf{x}_1 - \hat{\mathbf{x}}_1\|^2 = \|\mathbf{x}_1\|^2 + \|\hat{\mathbf{x}}_1\|^2 - 2\Re\left(\hat{\mathbf{x}}_1\mathbf{x}\right)$ we get

$$\begin{aligned}P(\underline{\mathbf{c}}_1 \to \hat{\underline{\mathbf{c}}}_1|\overline{\mathbf{H}}_1) &\leq P\left(\left(|h_{11}|+|h_{21}|\right)\left(\frac{3}{2\sqrt{2}}\|\mathbf{x}_1\|^2 - \frac{1}{2\sqrt{2}}\|\mathbf{x}_1-\hat{\mathbf{x}}_1\|^2 + \frac{1}{2\sqrt{2}}\|\hat{\mathbf{x}}_1\|^2\right)+\Re\left(\mathbf{z}_1^{\dagger}(\mathbf{x}_1-\hat{\mathbf{x}}_1)\right)\leq 0\right)\\ &= P\left(\kappa\left(|h_{11}|+|h_{21}|\right)+z_1^{'}\leq 0\right)\end{aligned}$$

where $\kappa = \frac{3}{2\sqrt{2}}\|\mathbf{x}_1\|^2 - \frac{1}{2\sqrt{2}}\|\mathbf{x}_1 - \hat{\mathbf{x}}_1\|^2 + \frac{1}{2\sqrt{2}}\|\hat{\mathbf{x}}_1\|^2$. $z_1^{'} = \Re\left(\mathbf{z}_1^{\dagger}\left(\mathbf{x}_1 - \hat{\mathbf{x}}_1\right)\right)$ is circularly symmetric complex while Gaussian noise of variance $\frac{N_0}{2}\|\mathbf{x} - \hat{\mathbf{x}}_1\|^2$. So the PEP is upperbounded as

$$P\left(\underline{\mathbf{c}}_1 \to \hat{\underline{\mathbf{c}}}_1|\overline{\mathbf{H}}_1\right) \leq P\left(\kappa\left(|h_{11}|+|h_{21}|\right)+z_1^{'}\leq 0\right) \tag{38}$$

As per ours notations, the decision variable γ as per (5) and (13) in [29] is given as

$$\gamma = \kappa\left(|h_{11}|+|h_{21}|\right)+z_1^{'} \tag{39}$$

So the probability of error which is given as $P\left(\gamma \leq 0\right)$ is given as

$$P\left(\underline{\mathbf{c}}_1 \to \hat{\underline{\mathbf{c}}}_1\right) = \frac{1}{2}\left\{1 - \frac{\sqrt{\rho_{11}\left(\rho_{11}+2\kappa\right)}+\sqrt{\rho_{21}\left(\rho_{21}+2\kappa\right)}}{\rho_{11}+\rho_{21}+2\kappa}\right\} \tag{40}$$

where $\rho_{ij} = \mathbb{E}\left(\left|h_{ij}\right|^2\right)/N_0$ is the SNR at the individual branch and κ is a constant that will depend on the constellation. (40) shows the full diversity order of 2, a result earlier derived for EGT in single-user MIMO systems in [14] using the approach of metrics of diversity order.

5.2. Multi-user MIMO

We now focus on the PEP of UE-1 in the multi-user MIMO mode as per system equation (3). Let $\mathbf{p}_1 = [1 \;\; q]^T$ where $q \in \{\pm 1, \; \pm j\}$. To have good channel separation between the UEs to be served in the multi-user MIMO mode [8], scheduling at the eNodeB would ensure $\mathbf{p}_2$ to be $[1 \;\; -q]^T$. The effective channel seen by the desired stream $x_{1,k}$ at UE-1 is given as $h_{1,k} = h^*_{11,k} + qh^*_{21,k}$ whereas the channel seen by the interference stream $x_{2,k}$ is $h_{2,k} = h^*_{11,k} - qh^*_{21,k}$. The max log MAP bit metric is written as

$$\Lambda_1^i\left(y_{1,k}, c_{k'}\right) \approx \min_{x_1 \in \chi^i_{1,c_{k'}}, x_2 \in \chi_2} \frac{1}{N_0}\left|y_{1,k} - \frac{1}{\sqrt{4}}h_{1,k}x_1 - \frac{1}{\sqrt{4}}h_{2,k}x_2\right|^2$$

Conditional PEP is given as

$$\begin{aligned} P\left(\underline{\mathbf{c}}_1 \to \hat{\underline{\mathbf{c}}}_1 | \overline{\mathbf{H}}_1\right) = P &\left(\sum_{k'} \min_{x_1 \in \chi^i_{1,c_{k'}}, x_2 \in \chi_2} \frac{1}{N_0}\left|y_{1,k} - \frac{1}{\sqrt{4}}h_{1,k}x_1 - \frac{1}{\sqrt{4}}h_{2,k}x_2\right|^2 \right. \\ &\left. \geq \sum_{k'} \min_{x_1 \in \chi^i_{1,\hat{c}_{k'}}, x_2 \in \chi_2} \frac{1}{N_0}\left|y_{1,k} - \frac{1}{\sqrt{4}}h_{1,k}x_1 - \frac{1}{\sqrt{4}}h_{2,k}x_2\right|^2\right) \end{aligned} \tag{41}$$

Let's denote

$$\begin{aligned} \tilde{x}_{1,k}, \tilde{x}_{2,k} &= \arg \min_{x_1 \in \chi^i_{1,c_{k'}}, x_2 \in \chi_2} \frac{1}{N_0}\left|y_{1,k} - \frac{1}{\sqrt{4}}h_{1,k}x_1 - \frac{1}{\sqrt{4}}h_{2,k}x_2\right|^2 \\ \hat{x}_{1,k}, \hat{x}_{2,k} &= \arg \min_{x_1 \in \chi^i_{1,\hat{c}_{k'}}, x_2 \in \chi_2} \frac{1}{N_0}\left|y_{1,k} - \frac{1}{\sqrt{4}}h_{1,k}x_1 - \frac{1}{\sqrt{4}}h_{2,k}x_2\right|^2 \end{aligned} \tag{42}$$

Note that

$$\left|y_{1,k} - \frac{1}{\sqrt{4}}\left(h_{1,k}x_{1,k} + h_{2,k}x_{2,k}\right)\right|^2 \geq \left|y_{1,k} - \frac{1}{\sqrt{4}}\left(h_{1,k}\tilde{x}_{1,k} + h_{2,k}\tilde{x}_{2,k}\right)\right|^2$$

So conditional PEP is given as

$$\begin{aligned} P\left(\underline{\mathbf{c}}_1 \to \hat{\underline{\mathbf{c}}}_1 | \overline{\mathbf{H}}_1\right) &\leq Q\left(\sqrt{\sum_{k,d_{free}} \frac{1}{8N_0}\left|h_{1,k}\left(x_{1,k} - \hat{x}_{1,k}\right) + h_{2,k}\left(x_{2,k} - \hat{x}_{2,k}\right)\right|^2}\right) \\ &= Q\left(\sqrt{\sum_{k,d_{free}} \frac{1}{8N_0}\left|\mathbf{h}_k^T\left(\mathbf{x}_k - \hat{\mathbf{x}}_k\right)\right|^2}\right) \end{aligned} \tag{43}$$

where $\mathbf{h}_k = \left[h^*_{11,k} \quad qh^*_{21,k} \quad h^*_{11,k} \quad -qh^*_{21,k}\right]^T$, $\mathbf{x}_k = \left[x_{1,k} \quad x_{1,k} \quad x_{2,k} \quad x_{2,k}\right]^T$ and $\hat{\mathbf{x}}_k = \left[\hat{x}_{1,k} \quad \hat{x}_{1,k} \quad \hat{x}_{2,k} \quad \hat{x}_{2,k}\right]^T$. We assume channel to be slow fading, i.e. the channel remains constant for the duration of one codeword. So the PEP can be written as

$$\begin{aligned} P\left(\underline{\mathbf{c}}_1 \rightarrow \hat{\underline{\mathbf{c}}}_1 | \overline{\mathbf{H}}_1\right) &\leq Q\left(\sqrt{\sum_{k,d_{free}} \frac{1}{8N_0}\left|\mathbf{h}^T\left(\mathbf{x}_k - \hat{\mathbf{x}}_k\right)\right|^2}\right) \\ &= Q\left(\sqrt{\frac{1}{8N_0}\mathbf{h}^\dagger \boldsymbol{\Delta}\boldsymbol{\Delta}^\dagger \mathbf{h}}\right) \end{aligned} \tag{44}$$

where $\boldsymbol{\Delta}\boldsymbol{\Delta}^\dagger$ is a 4×4 matrix while $\boldsymbol{\Delta}_{4\times d_{free}} = \left[\mathbf{x}_1 - \hat{\mathbf{x}}_1 \; \mathbf{x}_2 - \hat{\mathbf{x}}_2 \; \cdots \; \mathbf{x}_{k,d_{free}} - \hat{\mathbf{x}}_{k,d_{free}}\right]$. Using Chernoff bound, (44) is upper bounded by

$$P\left(\mathbf{c}_1 \rightarrow \hat{\mathbf{c}}_1 | \mathbf{h}\right) \leq \frac{1}{2}\exp\left(-\frac{1}{16N_0}\mathbf{h}^\dagger \boldsymbol{\Delta}\boldsymbol{\Delta}^\dagger \mathbf{h}\right) \tag{45}$$

The covariance matrix of the channel $\mathbf{h}$ is

$$E\left[\mathbf{h}\mathbf{h}^\dagger\right] = \mathbf{R} = \begin{bmatrix} 1 & 0 & 1 & 0 \\ 0 & 1 & 0 & -1 \\ 1 & 0 & 1 & 0 \\ 0 & -1 & 0 & 1 \end{bmatrix} \tag{46}$$

Its rank is two with its two identical eigenvalues being 2. Using the moment generating function of a Hermitian quadratic form in complex Gaussian random variable, we get

$$E_{\mathbf{h}}\left[\frac{1}{2}\exp\left(-\frac{1}{16N_0}\mathbf{h}^\dagger \boldsymbol{\Delta}\boldsymbol{\Delta}^\dagger \mathbf{h}\right)\right] \leq \frac{1}{2\det\left(\mathbf{I} + \frac{1}{16N_0}\mathbf{R}\boldsymbol{\Delta}\boldsymbol{\Delta}^\dagger\right)} \tag{47}$$

Note that the minimizations in (42) ensure that in $\boldsymbol{\Delta}$, $\hat{x}'_{1,k} - x'_{1,k}$ is always non-zero where $\hat{x}'_{2,k} - x'_{2,k}$ can be zero for $k = 1, \cdots, d_{free}$. So in the worst case scenario, $\boldsymbol{\Delta}$ would have only first two rows with non-zero elements. For the high SNR approximation, we get

$$P\left(\underline{\mathbf{c}}_1 \rightarrow \hat{\underline{\mathbf{c}}}_1\right) \leq \frac{1}{2}\left(\frac{16N_0}{\sigma^2}\right)^r \prod_{k=1}^{r} \frac{1}{\mu_k} \tag{48}$$

where r is the rank and μ_k are the eigenvalues of $\mathbf{R}\boldsymbol{\Delta}\boldsymbol{\Delta}^\dagger$. The minimum rank is one thereby indicating the diversity order of one. Note that as the derivation has involved Chernoff bound, so the exact PEP expression would involve some additional multiplicative factors but these factors will not affect the diversity order.

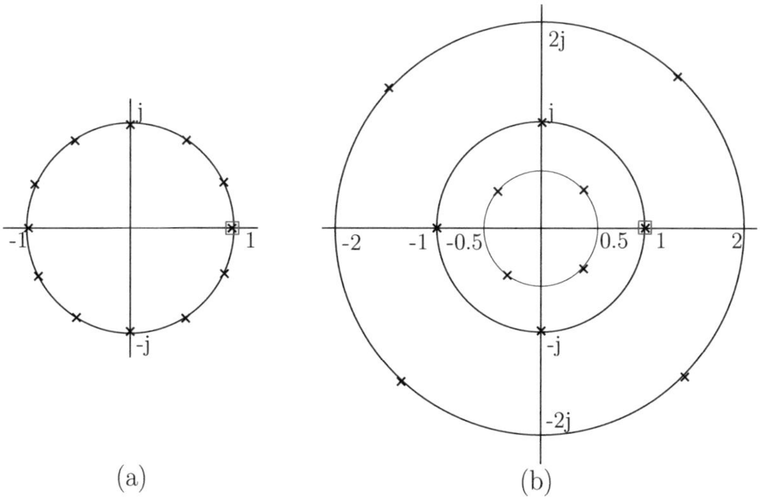

Figure 4. Two options of increasing the precoder codebook size. Fig.(a) corresponds to the option of increased angular resolution of LTE precoders while Fig.(b) corresponds to the option of enhanced levels of transmission. Square indicates the precoder entry for the first antenna while cross indicates the precoder entry for the second antenna.

6. The proposed feedback and codebook design

It was shown in the PEP analysis that the multi-user MIMO mode in LTE suffers from a loss of diversity. This loss is mainly attributed to the EGT characteristic of these precoders as will be shown in the next section. On the other hand, this transmission characteristic does not affect the diversity order in single-user MIMO mode. Focusing on this result, we propose a design of LTE precoders to offset this diversity loss.

LTE precoders are characterized by two features, i.e. angular resolution and EGT. Limited increase in the feedback can be either employed to increase the angular resolution of these structured precoders or it can be used to enhance the levels of transmission. Increasing the levels of transmission implies that additional feedback bits can be used to indicate an increase of the power level on either of the two antennas, i.e. creating more circles with different radii. For this we resorted to numerical optimization for fixing the radii of two circles and the precoders turn out to be $[1 \quad 2\exp(j\theta)]^T$ or $[2\exp(j\theta) \quad 1]^T$ where $\theta \in \{0, \pm 90°, 180°\}$. This approach gives 8 additional codebook entries, and 12 in total. Improving angular resolution is trivial, i.e. increasing equally angular spaced points on the unit circle but restricting to EGT, i.e. precoder is given as $[1 \quad \exp(j\theta)]^T$, where $\theta = 2\pi l/12, l = 0, \ldots, 11$. These two different codebook options have been illustrated in Fig. 4.

To quantize the proposed codebooks of size 12, $\lceil \log_2(12) \rceil = 4$ bits are needed. That means that we could add 4 more additional codebook entries for free, but it is not obvious how those extra entries should be designed in the case of the codebook with the additional transmission levels. On the other hand it can be argued that several PMI feedbacks (for example for different subbands) can be bundled to optimize the feedback rate.

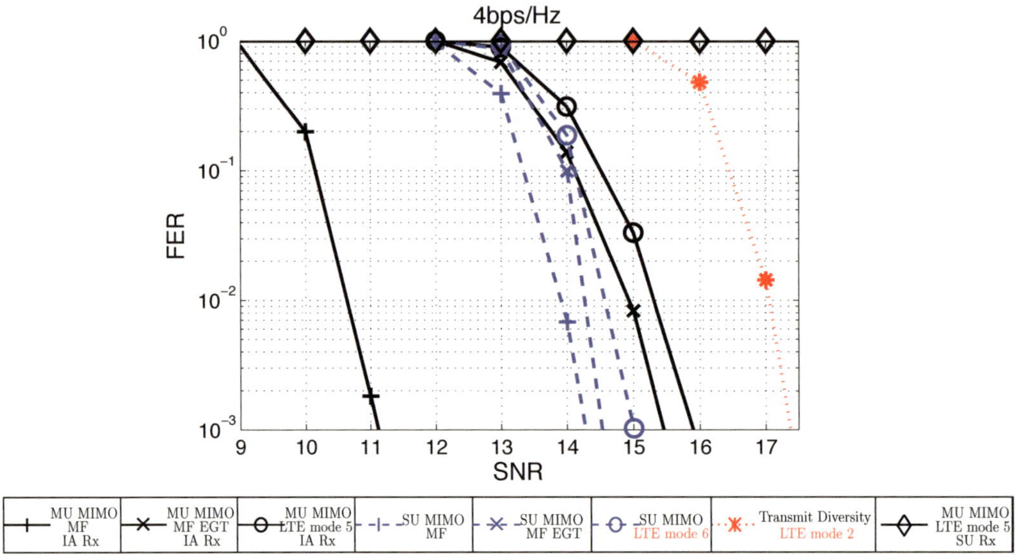

Figure 5. Downlink fast fading channel with the dual-antenna eNodeB and 2 single-antenna UEs. IA Rx indicates the low-complexity interference-aware receiver while SU Rx indicates the single-user receiver. MU MIMO and SU MIMO indicate multi-user and single-user MIMO respectively. To be fair in comparison amongst different schemes, sum rates are fixed, i.e. if 2 users are served with QPSK with rate 1/2 in the multi-user mode, then one user is served with QAM16 with rate 1/2 in the single-user mode thereby equating the sum rate in both cases to 2bps/Hz. 3GPP LTE rate 1/3 turbo code is used with different puncturing patterns.

7. Simulation results

Simulations are divided into 3 parts. In the first part, we look at the performance of the proposed interference-aware receiver structure for the multi-user MIMO mode in LTE while second part is dedicated to the sensitivity analysis of this receiver structure to the knowledge of the constellation of interference. This sensitivity analysis is motivated by the fact that the DCI formats in the transmission mode 5 (multi-user MIMO) do not include the information of the constellation of the co-scheduled UE. Third part looks at the diversity order of the EGT in both single-user and multi-user MIMO modes in LTE.

For the first part (Figs. 5 and 6), we consider the downlink of 3GPP LTE which is based on BICM OFDM transmission from the eNodeB equipped with two antennas using rate-1/3 LTE turbo code[1] [16] with rate matching to rate 1/2 and 1/4. We deliberate on both the cases of single and dual-antenna UEs. We consider an ideal OFDM system (no ISI) and analyze it in the frequency domain where the channel has iid Gaussian matrix entries with unit variance and is independently generated for each channel use. We assume no power control in the multi-user MIMO mode so two UEs have equal power distribution. Furthermore, all mappings of the coded bits to QAM symbols use Gray encoding. We focus on the FER while the frame length is fixed to 1056 information bits. As a reference, we consider the fall-back transmit diversity scheme (LTE mode 2 - Alamouti code) and compare it with the single-user and multi-user MIMO modes employing single-user receivers and

[1] The LTE turbo decoder design was performed using the coded modulation library www.iterativesolutions.com

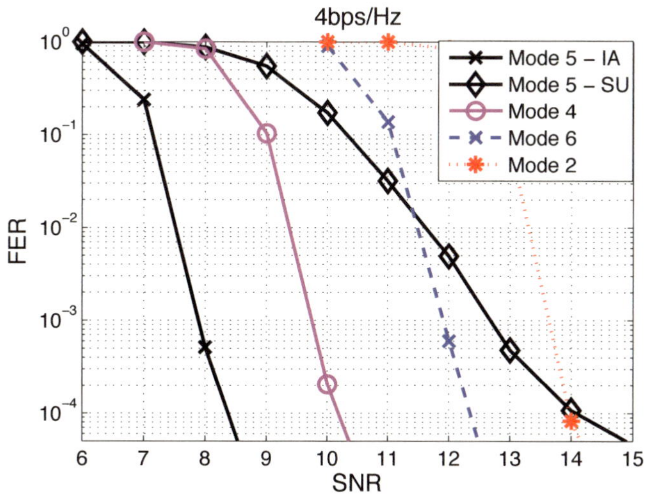

Figure 6. Downlink fast fading channel with the dual-antenna eNodeB and 2 dual-antenna UEs. IA indicates the low-complexity interference-aware receiver while SU indicates the single-user receiver. 3GPP LTE rate 1/3 turbo code is used with different puncturing patterns.

low-complexity interference-aware receivers. To analyze the degradation caused by the low-resolution and EGT of LTE precoders, we also look at the system performance employing the unquantized MF and unquantized MF EGT precoders. To be fair in the comparison of the LTE multi-user MIMO mode (mode 5) employing the geometric scheduling algorithm with the multi-user MIMO mode employing unquantized MF and MF EGT precoders, we consider the geometric scheduling algorithm (Section 4) based on the spatial angle between the two channels (equation (22)). Perfect CSIT is assumed for the case of MF and MF EGT precoding while error free feedback of 2 bits (PMI) to the eNodeB is assumed for LTE precoders. It is assumed that the UE has knowledge of the constellation of co-scheduled UE in the multi-user MIMO mode. It is further assumed that the UE knows its own channel from the eNodeB. So in multi-user MIMO mode, the UE can find the effective interference channel based on the fact that the eNodeB schedules the second UE on the same RE whose precoder is 180° out of phase of the precoder of the first UE. Fig. 5 shows the results for the case of single-antenna UEs. It shows enhanced performance of the multi-user MIMO mode once the UEs resort to intelligent detection by employing the low-complexity interference-aware receivers. The performance is severely degraded once the UEs resort to single-user detection. An interesting result is almost the equivalent performance of the unquantized MF EGT and low-resolution LTE precoders which shows that the loss with respect to the unquantized CSIT is attributed to the EGT rather than the low-resolution of LTE precoders.

Fig. 6 shows the results for the case of dual-antenna UEs and focuses on different LTE modes employing LTE precoders. It shows the degraded performance of single-user detection which is due to the fact that the rate with single-user detection gets saturated at high SNR due to the increased interference strength as was shown in Section. 4. So the performance of single-user detection is degraded as the spectral efficiency is higher than the rate or mutual information of the single-user detection. For single-user MIMO (Mode 6), there is no saturation of the rate at high SNR as there is no interference. So mode 6 performs better than mode 5 at

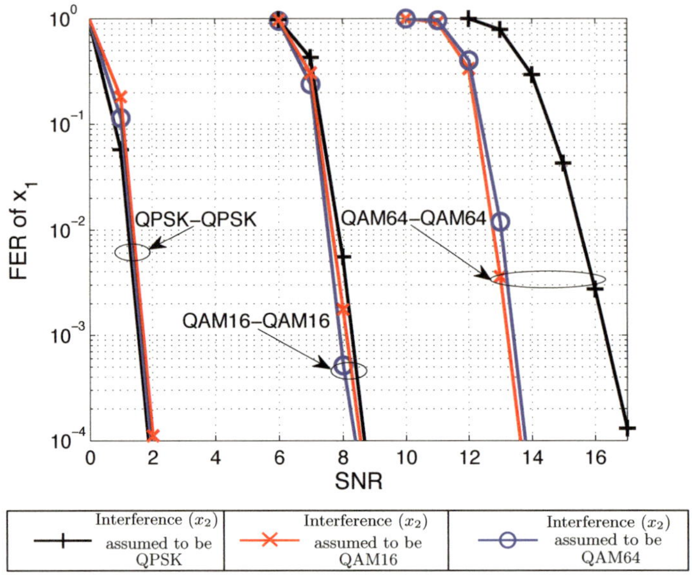

Figure 7. Interference sensitivity for the multi-user MIMO mode in LTE. Three sets of simulations are shown. QPSK-QPSK indicates that both x_1 and x_2 belong to QPSK. UE-1 does not know the constellation of interference (x_2) and assumes it to be QPSK, QAM16 and QAM64.

high SNR once UEs employ single-user detection. However if UEs resort to the intelligent interference-aware detection, the multi-user MIMO mode shows enhanced performance over other transmission modes in LTE. No degradation of LTE multi-user MIMO mode is observed at higher spectral efficiencies once UEs have receive diversity (dual antennas).

In the second part of simulations, we look at the sensitivity of the proposed receiver structure to the knowledge of the constellation of co-scheduled UE for the multi-user MIMO mode in LTE. The simulation settings are same as of the first part except that we consider the case when UE has no knowledge of the constellation of co-scheduled UE. The UE assumes this unknown interference constellation to be QPSK, QAM16 or QAM64 and the results for these different assumptions are shown in Fig. 7. Results show that there is negligible degradation in the performance of the proposed receiver if the interfering constellation is assumed to be QAM16 or QAM64. However, there is significant degradation if the interference is assumed to be QPSK when it actually comes from QAM64. It indicates that assuming interference to be from a higher order modulation amongst the possible modulation alphabets leads to the best compromise as this assumption includes the lower modulation orders as special cases (with proper scaling). However the converse is not true, i.e. assuming interference from lower modulation order cannot include higher order modulations. As LTE and LTE-Advanced restrict the transmission to three modulations (QPSK, QAM16 and QAM64), assuming interference to be QAM64 (or even QAM16) leads to better performance. If the interference constellation also includes QAM256, then assuming interference to be QAM256 (or even QAM64) would lead to better results. These results have not been shown here as LTE and LTE-Advanced do not support QAM256 modulation. The proposed receiver structure, therefore, can still exploit the discrete nature of the interference even if it does not know

http://dx.doi.org/10.5772/57134

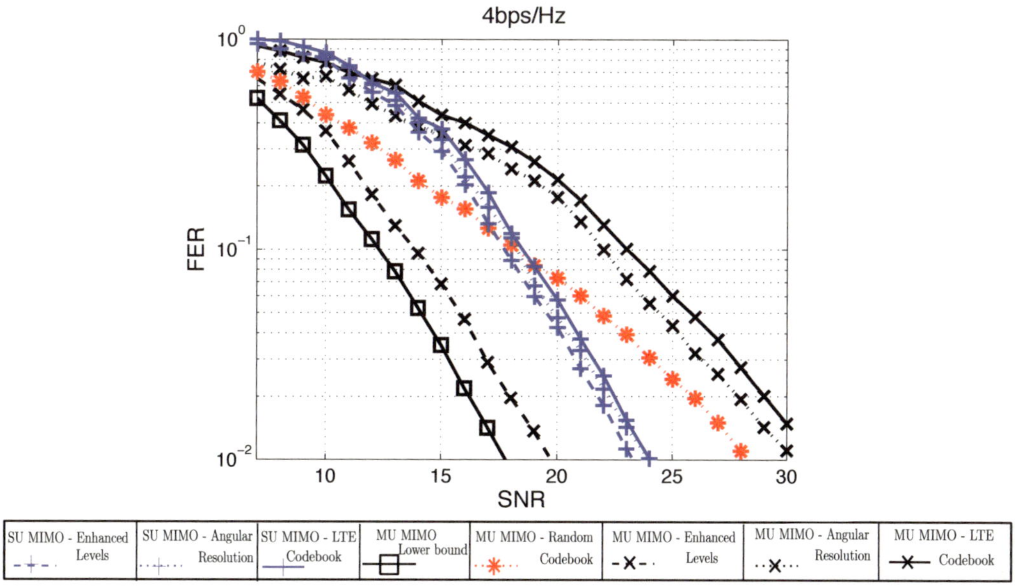

Figure 8. Proposed precoder codebook. Downlink channel with dual-antenna eNodeB and two single-antenna UEs. The figure illustrates the performance for the sum rate of 4bps/Hz. SNR is the transmit SNR while sum rate is same for single-user and multi-user MIMO, i.e. if two UEs are served with QPSK, rate 1/2, in multi-user mode, then one UE is served with QAM16, rate 1/2, in the single-user mode. SU MIMO and MU MIMO indicate single-user and multi-user MIMO.

its modulation order. As the complexity of this receiver structure is independent of the constellation of interference, the assumption of higher order modulation does not add to the complexity of detection.

In the third set of simulations, we look at the diversity order of the single-user MIMO and multi-user MIMO schemes in LTE. The system settings are same as in the first part but now we consider slow fading environment, i.e. the channel remains constant for the duration of one codeword. Fig. 8 shows significant improvement in the performance of the multi-user MIMO mode when additional codebook entries are employed to increase the levels of transmission as compared to the case of increasing the angular resolution of precoders. However creating two levels of transmission leads to significant improvement as the performance moves closer to the upper bound. This hypothetical upper bound is the performance curve for MF precoder in multi-user MIMO mode without any interference, i.e. the eNodeB serves two UEs with their respective MF based precoders and the two UEs do not see any interference. The change of the slope of FER curve with increased levels of transmission indicates improved diversity as compared to the case of increased angular resolution. On the other hand, little gain is observed in the single-user mode (LTE transmission mode 6) with additional codebook entries which is expected as the standard LTE precoders have been optimized for the single-user transmission [22]. For comparison purposes, we have also considered the case of random codebooks. The main advantage of random codebooks is that they indicate some sort of performance lower bound and with any intelligent feedback design, system is bound to perform better.

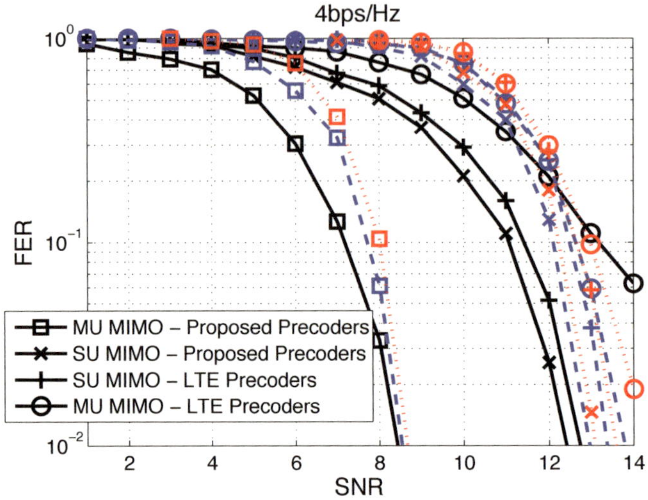

Figure 9. Performance of the proposed precoder codebook in 3GPP LTE channel models [18]. Black continuous lines show the Extended Pedestrian A model (EPA), blue dashed lines show Extended Vehicular A model (EVA), while red dotted lines show Extended Typical Urban model (ETU).

Fig. 9 shows the case where we have considered 3GPP LTE channel model introduced in [18] for three representative scenarios, i.e. pedestrian, vehicular and typical urban scenario. The transmission chain is dominantly LTE compliant with 15 KHz subcarrier-spacing and 20 MHz system bandwidth. The results confirm the earlier findings of the improved performance of proposed codebook design (enhanced levels of transmission) for multi-user transmission mode. Pedestrian channel offers less diversity in the channel as compared to the vehicular channel, so the performance of LTE precoders for multi-user MIMO in severely degraded in the former case. However as the proposed precoder design recovers the lost order of diversity, there is an improvement of 6dB at the target FER of 10^{-1}.

8. Conclusions

In this chapter, we have looked at the feasibility of the multi-user MIMO for future wireless systems which are characterized by low-level quantization of CSIT. We have shown that multi-user MIMO can deliver its promised gains if the UEs resort to intelligent detection rather than the sub-optimal single-user detection. To this end, we have proposed a low-complexity interference-aware receiver structure which is characterized by the exploitation of the structure of residual interference. We have further investigated the impact of low-level fixed rate feedback on the performance of multi-user MIMO in LTE systems. We have analyzed two important characteristics of the LTE precoders, i.e. low resolution and EGT. We have shown that the performance loss of the LTE precoders in the multi-user MIMO mode is attributed to their characteristic of EGT rather than their low resolution. We have proposed a feedback and precoding design and have shown that the performance in multi-user MIMO significantly improves once strategy of more levels of transmission is resorted to as compared to the case of increased angular resolution. The work presented in this chapter is not merely confined to the framework of LTE, rather it gives the receiver structure and precoding design guidelines for modern wireless systems.

Appendix A
Mutual information for finite alphabets

The mutual information for UE-1 for finite size QAM constellation with $|\chi_1| = M_1$ takes the form as

$$I\left(Y_1; X_1|\mathbf{h}_1^{\dagger}, \mathbf{P}\right) = \mathcal{H}\left(X_1|\mathbf{h}_1^{\dagger}, \mathbf{P}\right) - \mathcal{H}\left(X_1|Y_1, \mathbf{h}_1^{\dagger}, \mathbf{P}\right)$$
$$= \log M_1 - \mathcal{H}\left(X_1|Y_1, \mathbf{h}_1^{\dagger}, \mathbf{P}\right) \tag{49}$$

where $\mathcal{H}(.) = -E \log p(.)$ is the entropy function. The second term of (49) is given as

$$\mathcal{H}\left(X_1|Y_1, \mathbf{h}_1^{\dagger}, \mathbf{P}\right) = \sum_{x_1}\int_{y_1}\int_{\mathbf{h}_1^{\dagger}\mathbf{p}_1}\int_{\mathbf{h}_1^{\dagger}\mathbf{p}_2} p\left(x_1, y_1, \mathbf{h}_1^{\dagger}\mathbf{p}_1, \mathbf{h}_1^{\dagger}\mathbf{p}_2\right) \log \frac{1}{p\left(x_1|y_1, \mathbf{h}_1^{\dagger}\mathbf{p}_1, \mathbf{h}_1^{\dagger}\mathbf{p}_2\right)} dy_1 d(\mathbf{h}_1^{\dagger}\mathbf{p}_1) d(\mathbf{h}_1^{\dagger}\mathbf{p}_2)$$
$$= \sum_{x_1}\sum_{x_2}\int_{y_1}\int_{\mathbf{h}_1^{\dagger}\mathbf{p}_1}\int_{\mathbf{h}_1^{\dagger}\mathbf{p}_2} p\left(x_1, x_2, y_1, \mathbf{h}_1^{\dagger}\mathbf{p}_1, \mathbf{h}_1^{\dagger}\mathbf{p}_2\right) \log \frac{\sum_{x_1'}\sum_{x_2'} p\left(y_1|x_1', x_2', \mathbf{h}_1^{\dagger}\mathbf{p}_1, \mathbf{h}_1^{\dagger}\mathbf{p}_2\right)}{\sum_{x_2'} p\left(y_1|x_1, x_2', \mathbf{h}_1^{\dagger}\mathbf{p}_1, \mathbf{h}_1^{\dagger}\mathbf{p}_2\right)} dy_1 d(\mathbf{h}_1^{\dagger}\mathbf{p}_1) d(\mathbf{h}_1^{\dagger}\mathbf{p}_2) \tag{50}$$

where $x_1' \in \chi_1$ and $x_2' \in \chi_2$. Conditioned on the channel and the precoder, there is one source of randomness, i.e. noise. So (50) can be extended as

$$\mathcal{H}\left(X_1|Y_1, \mathbf{h}_1^{\dagger}, \mathbf{P}\right) = \frac{1}{M_1 M_2}\sum_{\mathbf{x}} E_{z_1} \log \frac{\sum_{\mathbf{x}'} \exp\left[-\frac{1}{N_0}\left|\mathbf{h}_1^{\dagger}\mathbf{p}_1 x_1 + \mathbf{h}_1^{\dagger}\mathbf{p}_2 x_2 + z_1 - \mathbf{h}_1^{\dagger}\mathbf{p}_1 x_1' - \mathbf{h}_1^{\dagger}\mathbf{p}_2 x_2'\right|^2\right]}{\sum_{x_2'} \exp\left[-\frac{1}{N_0}\left|\mathbf{h}_1^{\dagger}\mathbf{p}_2 x_2 + z_1 - \mathbf{h}_1^{\dagger}\mathbf{p}_2 x_2'\right|^2\right]}$$
$$= \frac{1}{M_1 M_2}\sum_{\mathbf{x}} E_{z_1} \log \frac{\sum_{\mathbf{x}'} \exp\left[-\frac{1}{N_0}\left|\mathbf{h}_1^{\dagger}\mathbf{P}\left(\mathbf{x} - \mathbf{x}'\right) + z_1\right|^2\right]}{\sum_{x_2'} \exp\left[-\frac{1}{N_0}\left|\mathbf{h}_1^{\dagger}\mathbf{P}\left(\mathbf{x} - \mathbf{x}_2'\right) + z_1\right|^2\right]} \tag{51}$$

where $M_2 = |\chi_2|$, $\mathbf{x} = [x_1\ x_2]^T$, $\mathbf{x}' = \left[x_1'\ x_2'\right]^T$ and $\mathbf{x}_2' = \left[x_1\ x_2'\right]^T$. The mutual information for UE-1 can be rewritten as

$$I\left(Y_1; X_1|\mathbf{h}_1^{\dagger}, \mathbf{P}\right) = \log M_1 - \frac{1}{M_1 M_2}\sum_{\mathbf{x}} E_{z_1} \log \frac{\sum_{\mathbf{x}'} p\left(y_1|\mathbf{x}', \mathbf{h}_1^{\dagger}, \mathbf{P}\right)}{\sum_{x_2'} p\left(y_1|\mathbf{x}_2', \mathbf{h}_1^{\dagger}, \mathbf{P}\right)} \tag{52}$$

The above quantities can be easily approximated using sampling (Monte-Carlo) methods with N_z realizations of noise and N_{h_1} realizations of the channel $\mathbf{h}_1^{\dagger}$ where the precoding matrix depends on the channel. So we can rewrite (52) as (53)

$$I\left(Y;X_1|\mathbf{h}_1^\dagger\mathbf{p}_1,\mathbf{h}_1^\dagger\mathbf{p}_2\right)=\log M_1-\frac{1}{M_1M_2N_zN_{h_1}}\sum_{x_1}\sum_{x_2}\sum_{\mathbf{h}_1}^{N_{h_1}}\sum_{z_1}^{N_z}\log\frac{\sum_{x_1'}\sum_{x_2'}\exp\left[-\frac{1}{N_0}\left|y_1-\mathbf{h}_1^\dagger\mathbf{p}_1x_1'-\mathbf{h}_1^\dagger\mathbf{p}_2x_2'\right|^2\right]}{\sum_{x_2'}\exp\left[-\frac{1}{N_0}\left|y_1-\mathbf{h}_1^\dagger\mathbf{p}_1x_1-\mathbf{h}_1^\dagger\mathbf{p}_2x_2'\right|^2\right]}$$

$$=\log M_1-\frac{1}{M_1M_2N_zN_{h_1}}\sum_{x_1}\sum_{x_2}\sum_{\mathbf{h}_1}^{N_{h_1}}\sum_{z_1}^{N_z}\log\frac{\sum_{x_1'}\sum_{x_2'}\exp\left[-\frac{1}{N_0}\left|\mathbf{h}_1^\dagger\mathbf{p}_1x_1+\mathbf{h}_1^\dagger\mathbf{p}_2x_2+z_1-\mathbf{h}_1^\dagger\mathbf{p}_1x_1'-\mathbf{h}_1^\dagger\mathbf{p}_2x_2'\right|^2\right]}{\sum_{x_2'}\exp\left[-\frac{1}{N_0}\left|\mathbf{h}_1^\dagger\mathbf{p}_2x_2+z_1-\mathbf{h}_1^\dagger\mathbf{p}_2x_2'\right|^2\right]} \quad (53)$$

Similarly the mutual information for UE-2 is given as

$$I\left(Y_2;X_2|\mathbf{h}_2^\dagger,\mathbf{P}\right)=\log M_2-\frac{1}{M_1M_2}\sum_{\mathbf{x}}E_{z_2}\log\frac{\sum_{\mathbf{x}'}p\left(y_2|\mathbf{x}',\mathbf{h}_2^\dagger,\mathbf{P}\right)}{\sum_{x_1'}p\left(y_2|\mathbf{x}_1',\mathbf{h}_2^\dagger,\mathbf{P}\right)} \quad (54)$$

where $\mathbf{x}_1'=\left[x_1'\ x_2\right]^T$.

For the case of single-user MIMO mode, the mutual information is given by

$$I\left(Y_1;X_1|\mathbf{h}_1^\dagger,\mathbf{p}_1\right)=\log M_1-\mathcal{H}\left(X_1|Y_1,\mathbf{h}_1^\dagger,\mathbf{p}_1\right) \quad (55)$$

where the second term is given by

$$\mathcal{H}\left(X_1|Y_1,\mathbf{h}_1^\dagger,\mathbf{p}_1\right)=\sum_{x_1}\int_{y_1}\int_{\mathbf{h}_1^\dagger\mathbf{p}_1}p\left(x_1,y_1,\mathbf{h}_1^\dagger\mathbf{p}_1\right)\log\frac{1}{p\left(x_1|y_1,\mathbf{h}_1^\dagger\mathbf{p}_1\right)}dy_1d(\mathbf{h}_1^\dagger\mathbf{p}_1)$$

$$=\sum_{x_1}\int_{y_1}\int_{\mathbf{h}_1^\dagger\mathbf{p}_1}p\left(x_1,y_1,\mathbf{h}_1^\dagger\mathbf{p}_1\right)\log\frac{\sum_{x_1'}p\left(y_1|x_1',\mathbf{h}_1^\dagger\mathbf{p}_1\right)}{p\left(y_1|x_1,\mathbf{h}_1^\dagger\mathbf{p}_1\right)}dy_1d(\mathbf{h}_1^\dagger\mathbf{p}_1)$$

$$=\frac{1}{M_1N_zN_{h_1}}\sum_{x_1}\sum_{\mathbf{h}_1^\dagger}^{N_{h_1}}\sum_{z_1}^{N_z}\log\frac{\sum_{x_1'}\exp\left[-\frac{1}{N_0}\left|y_1-\mathbf{h}_1^\dagger\mathbf{p}_1x_1'\right|^2\right]}{\exp\left[-\frac{1}{N_0}\left|y_1-\mathbf{h}_1^\dagger\mathbf{p}_1x_1\right|^2\right]} \quad (56)$$

where N_{h_1} are the number of channel realizations of the channel $\mathbf{h}_1^\dagger$. Note that the precoding vector $\mathbf{p}_1$ is dependent on the channel $\mathbf{h}_1^\dagger$.

Author details

Rizwan Ghaffar[1,*], Raymond Knopp[2] and Florian Kaltenberger[2]

* Address all correspondence to: rizwan.ghaffar@eurecom.fr

1 Samsung Research America, USA
2 EURECOM, France

http://dx.doi.org/10.5772/57134

References

[1] Alamouti, S. [1998]. A simple transmit diversity technique for wireless communications, *IEEE Journal on Selected Areas in Communications* 16(8): 1451–1458.

[2] Au-Yeung, C. K. & Love, D. J. [2007]. On the performance of random vector quantization limited feedback beamforming in a MISO system, *IEEE Transactions on Wireless Communications* 6(2): 458 –462.

[3] Brunel, L. [2004]. Multiuser detection techniques using maximum likelihood sphere decoding in multicarrier CDMA systems, *IEEE Transactions on Wireless Communications* 3(3): 949 – 957.

[4] Caire, G., Taricco, G. & Biglieri, E. [1998]. Bit-interleaved Coded Modulation, *IEEE Transactions on Information Theory* 44(3): 927–946.

[5] Choi, J. W., Singer, A., Lee, J. & Cho, N. I. [2010]. Improved linear soft-input soft-output detection via soft feedback successive interference cancellation, *IEEE Transactions on Communications* 58(3): 986 –996.

[6] Gesbert, D., Kountouris, M., Heath, R., Chae, C.-B. & Salzer, T. [2007]. Shifting the MIMO paradigm, *IEEE Signal Processing Magazine* 24(5): 36 –46.

[7] Ghaffar, R. & Knopp, R. [2010]. Making Multiuser MIMO work for LTE, *IEEE 21-st International Symposium on Personal, Indoor and Mobile Radio Communications (PIMRC 2010)*, Istanbul.

[8] Ghaffar, R. & Knopp, R. [2011]. Interference-Aware Receiver Structure for Multi-User MIMO and LTE, *EURASIP Journal on Wireless Communications and Networking* .

[9] Ghaffar, R. & Knopp, R. [2012]. Interference Suppression Strategy for Cell-Edge Users in the Downlink, *IEEE Transactions on Wireless Communications* 11(1): 154 –165.

[10] Gradshteyn, I. & Ryzhik, I. [2000]. *Table of Integrals, Series, and Products*, Academic Press, San Diego, USA.

[11] Jindal, N. [2006]. MIMO broadcast channels with finite rate feedback, *IEEE Transactions on Information Theory* 52: 5045âĂŞ–5060.

[12] Lamare, R. d. & Sampaio-Neto, R. [2008]. Minimum Mean-Squared Error Iterative Successive Parallel Arbitrated Decision Feedback Detectors for DS-CDMA Systems, *IEEE Transactions on Communications* 56(5): 778 –789.

[13] Li, X., Chindapol, A. & Ritcey, J. [2002]. Bit-interleaved coded modulation with iterative decoding and 8 PSK signaling, *IEEE Transactions on Communications* 50(8): 1250 – 1257.

[14] Love, D. & Heath Jr., R. [2003]. Equal gain transmission in multiple-input multiple-output wireless systems, *IEEE Transactions on Communications* 51(7): 1102 – 1110.

[15] Love, D., Heath, R., Lau, V., Gesbert, D., Rao, B. & Andrews, M. [2008]. An overview of limited feedback in wireless communication systems, *IEEE Journal on Selected Areas in Communications* 26(8): 1341 –1365.

[16] LTE [2009a]. *Evolved Universal Terrestrial Radio Access (E-UTRA); Channel Coding and Multiplexing, Release 8, V.8.6.0*, 3GPP TS 36.212.

[17] LTE [2009b]. *Evolved Universal Terrestrial Radio Access (E-UTRA); Physical Layer Procedures, Release 10, V.8.6.0*, 3GPP TS 36.213.

[18] LTE [June 2011]. *Evolved Universal Terrestrial Radio Access (E-UTRA); User Equipment (UE) radio transmission and reception*, 3GPP TS 36.101 Version 10.3.0 Release 10.

[19] LTE-A [2011]. *Evolved Universal Terrestrial Radio Access (E-UTRA); Physical Layer Procedures, Release 10, V.10.1.0*, 3GPP TS 36.213.

[20] Lupas, R. & Verdu, S. [1989]. Linear multiuser detectors for synchronous code-division multiple-access channels, *IEEE Transactions on Information Theory* 35(1): 123 –136.

[21] Qi, X., Alouini, M.-S. & Ko, Y.-C. [2003]. Closed-form analysis of dual-diversity equal-gain combining over Rayleigh fading channels, *IEEE Transactions on Wireless Communication* 2(6): 1120 – 1125.

[22] Sesia, S., Toufik, I. & Baker, M. [2009]. *LTE, The UMTS Long Term Evolution: From Theory to Practice*, Wiley.

[23] Speth, M., Jansen, A. & Meyr, H. [2000]. Iterative multiuser detection for bit interleaved coded modulation, *IEEE International Conference on Communications, ICC 2000*, Vol. 2, pp. 894 –898 vol.2.

[24] Telatar, I. E. [1999]. Capacity of multiâĂŞantenna Gaussian channels, *European Transactions on Telecommunications* 10(6): 585–595.

[25] Verdu, S. [1998]. *Multiuser Detection*, Cambridge University Press, U.K.

[26] Wang, X. & Poor, H. [1999]. Iterative (turbo) soft interference cancellation and decoding for coded CDMA, *IEEE Transactions on Communications* 47(7): 1046 –1061.

[27] Zakhour, R., Ho, Z. K. M. & Gesbert, D. [2009]. Distributed beamforming coordination in multicellular MIMO systems, *IEEE 69th Vehicular Technology Conference,VTC-Spring. April, 26-29, 2009, Barcelona, Spain*.

[28] Zarikoff, B., Cavers, J. & Bavarian, S. [2007]. An iterative groupwise multiuser detector for overloaded MIMO applications, *IEEE Transactions on Wireless Communications* 6(2): 443 –447.

[29] Zhang, Q. [1997]. Probability of error for equal-gain combiners over Rayleigh channels: some closed-form solutions, *IEEE Transactions on Communications* 45(3): 270 –273.

Section 4

Interference Cancellation in Multiuser MIMO Systems

Chapter 4

Multi-User Interference Suppression by Using Frequency Domain Adaptive Antenna Array

Wei Peng and Fumiyuki Adachi

Additional information is available at the end of the chapter

http://dx.doi.org/10.5772/57132

1. Introduction

The target data rate for the next generation wireless communication network will be around 1Gbps. To realize such a high data rate transmission, broadband transmission has been used in the current wireless communication network and it is also going to be employed by the next generation network. Due to the multi-path fading with large delay spread, broadband wireless channel is characterized by severe frequency selectivity [1]. As a result, it is necessary to suppress the inter-symbol interference (ISI) at the receiver. The ISI can be suppressed by time domain equalization techniques such as maximum likelihood sequence estimation (MLSE) [2]. However, when the data rate increases, the number of resolvable propagation paths increases as well and hence, the complexity of MLSE grows exponentially to the number of paths. Fortunately, the ISI problem can be solved by introducing frequency domain equalization (FDE) [3] at the receiver. It is well known that the frequency selectivity problem can be solved by the use of multi-carrier transmission technique such as orthogonal frequency division multiple access (OFDMA) [4] for the downlink (from base station (BS) to mobile users) transmission. However, the multi-carrier transceivers are suffering from high peak-to-average power ratio (PAPR) problem which can lead to severe performance degradation. To solve the high PAPR problem, conventional single-carrier (SC) transmission, again, attracted much interest. Recently, the combination of SC-FDE and frequency division multiple access (called SC-FDMA) [5] has been considered as a more suitable solution for the uplink (from mobile users to BS) transmission. On the other hand, in order to save the bandwidth usage, the same carrier frequency/frequencies may be reused by neighboring cells to increase the bandwidth efficiency. As a result, co-channel interference (CCI) [6] becomes the dominant performance limitation instead of the thermal noise. In addition, multi-user interference (MUI) occurs when multiple users transmit simultaneously within the same cell (the MUI and CCI together is called multi-access interference (MAI)). Therefore, interference cancellation is necessary in uplink transmissions.

Recently, distributed antenna network (DAN) [7] has been proposed to solve the transmit power problem in broadband signal transmissions. As the data rate increases, impractically large transmit power will be required to realize the high data rate if cell coverage is kept unchanged. Otherwise, the cell coverage has to be reduced if the transmit power is kept unchanged. DAN was proposed as a solution to increase the cell coverage while maintaining the low transmit power. In the DAN, a number of antennas are distributed in each cell and those antennas are connected with the DAN central processor (which is similar to the BS in conventional cellular system) through optical cables. A mobile user can communicate with its' nearby located antennas even when it is at the cell edge. Therefore, the transmit power in DAN can be kept low while the coverage of the cell can be greatly increased.

In the previous studies [8, 9], a SC frequency domain adaptive antenna array (SC-FDAAA) for the uplink transmission has been proposed and it has been shown that the SC-FDAAA can effectively suppress MAI in a severely frequency selective fading channel. In this article, we will present the performance of DAN SC-FDAAA and compare SC-FDAAA in DAN and in conventional cellular system with centralized antennas at the BS (referred to as CAN system hereafter).

The rest of the article is organized as follows. The system model is given in Section II. SC-FDAAA for DAN and CAN will be described in Section III. The post SC-FDAAA signal to interference plus noise (SINR) will be given in Section IV. The performance of SC-FDAAA will be shown in Section V, both bit error rate (BER) distribution and the system capacity will be presented. Finally, the article will be concluded by Section VI.

2. System model

In cellular system, the same carrier frequency/frequencies will be reused in neighboring cells to effectively utilize the limited spectrum. Cellular structures with frequency reuse factors (FRFs) of 1, 3, 4 and 7 are shown in Fig. 1 as examples of frequency reuse. And FRF=1, 3, 4, 7, 9 and 12 will be considered in this study. The commonly used first layer CCI model is used here, i.e., only the CCI from the first layer neighboring cells will be considered and the number of CCI cells will be B =6. As stated in the Introduction, we are going to use both CAN and DAN in each cell. CAN and DAN structures are shown in Fig. 2. There are totally N_r centralized / distributed antennas. In the DAN system, those antennas are connected to the DAN central processor by optical fibers. The received signals will be transmitted to the DAN central processor and in order to lower the cost, signal processing will be carried out by the DAN central processor.

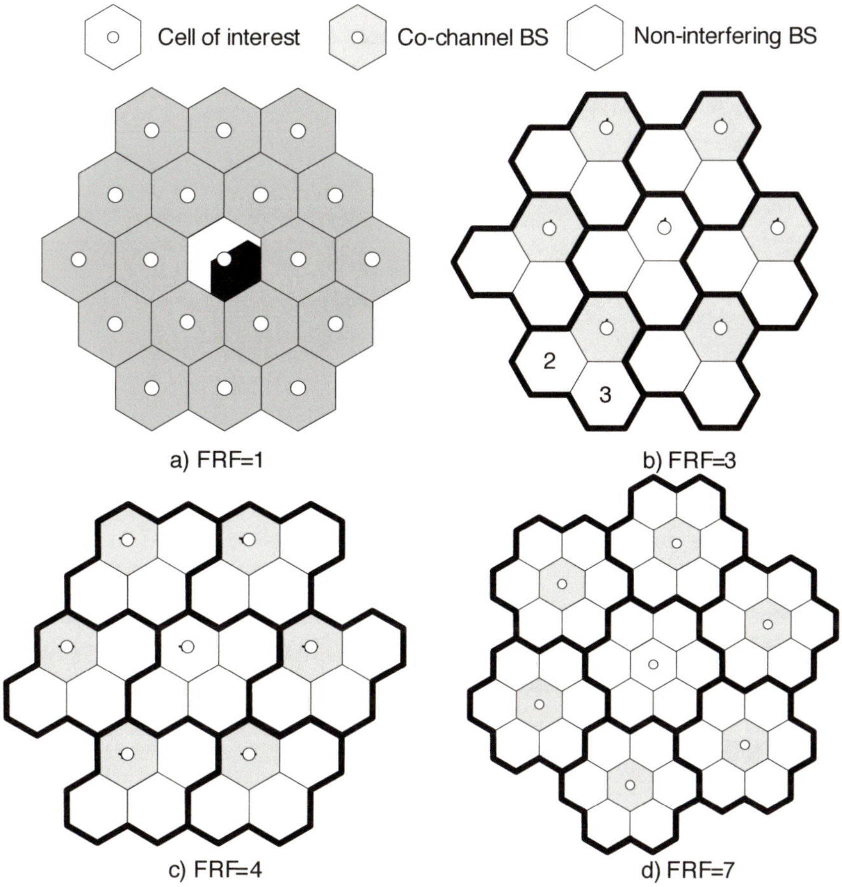

Figure 1. Frequency reuse in cellular system.

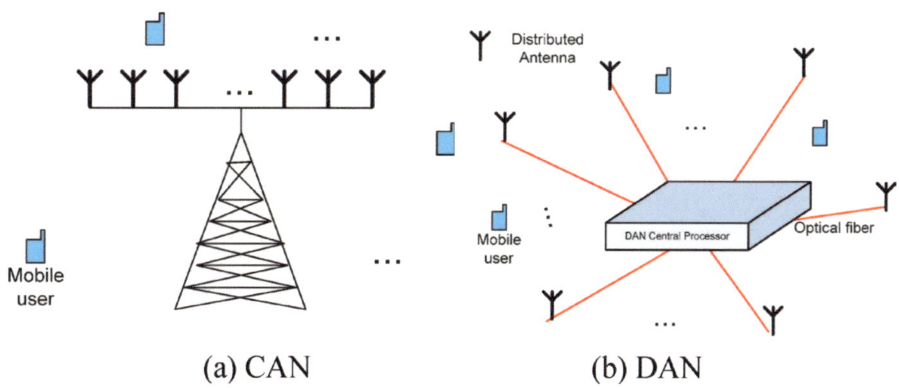

Figure 2. CAN system and DAN system.

It is assumed that there are U users within each cell and each user is equipped with one omni antenna. A block fading channel between each user and each antenna is assumed, i.e., the channel remains unchanged during the transmission period of a block. In this article, the symbol-spaced discrete time representation of the signal is used.

Assuming an L − path channel, the impulse response of the channel between the uth user and the mth antenna can be expressed as

$$h_{u,m}(\tau) = \sum_{l=0}^{L-1} h_{u,m,l}\delta(\tau - \tau_l) \tag{1}$$

where $h_{u,m,l}$ and τ_l are the path gain and time delay of the lth path, respectively. $h_{u,m,l}$ follows complex Gaussian distribution and satisfies $\sum_{l=0}^{L-1} E\{|h_{u,m,l}|^2\}=1$, where $E\{\cdot\}$ represents the expectation. It is assumed that the time delay is a multiple integer of the symbol duration and $\tau_l = l$ is used. The cyclic-prefixed (CP) block signal transmission is used to make the received symbol block to be a circular convolution of the transmitted symbol block and the channel impulse response as well as to avoid inter block interference (IBI). It is also assumed that the CP is longer than the maximum path delay of the signal. In the following, we omit the insertion and removal of the CP for the simplicity.

The baseband equivalent received signal block $\{r_m(t); t=0\sim N_c\}$ of N_c symbols at the mth antenna is given by

$$\begin{aligned} r_m(t) = {} & \sqrt{P_0\delta_{0,m}^{-\alpha}} \sum_{l=0}^{L-1} h_{0,m,l}s_0(t-l) + \sum_{u=1}^{U-1} \sqrt{P_u\delta_{u,m}^{-\alpha}} \sum_{l=0}^{L-1} h_{u,m,l}s_u(t-l) \\ & + \sum_{i=1}^{B} \sum_{u_i=0}^{U_i-1} \sqrt{P_{i,u_i}\delta_{i,u_i,m}^{-\alpha}} \sum_{l=0}^{L-1} h_{u_i,m,l}s_{u_i}(t-l) + n_m(t), \end{aligned} \tag{2}$$

where $s_u(t)$ and P_u are respectively the transmit signal and transmit signal power of the uth user $(u=0\sim U-1)$; s_{u_i} and P_{i,u_i} are respectively the transmit signal and transmit signal power of the $u_i{}^{th}$ user in the ith co-channel cell; $\delta_{0,m}$ represents the distance between the desired user and the mth antenna; $\delta_{i,m}$ represents the distance between the ith interfering user and the mth antenna; $\delta_{i,u_i,m}$ and $h_{u_i,m,l}$ are respectively the distance and channel gain between the CCI user and the mth antenna; α represents the path loss exponent in dB; and $n_m(t)$ is the additive white Gaussian noise (AWGN). To simplify the analysis, no shadowing loss is assumed.

Let the transmit signal from the $u=0^{th}$ user be the desired signal and the transmit signals from the other users be the interfering signals. The frequency domain representation of (2) is given by

$$R_m(k)=H_{0,m}(k)S_0(k)+\sum_{u=1}^{U-1}H_{u,m}(k)S_u(k)+\sum_{i=1}^{B}\sum_{u_i=0}^{U_i-1}H_{u_i,m}(k)S_{i,u_i}(k)+N_m(k), \tag{3}$$

where

$$\begin{cases} S_u(k)=\dfrac{1}{\sqrt{N_c}}\sqrt{P_u\delta_{u,m}^{-\alpha}}\displaystyle\sum_{t=0}^{N_c-1}s_u(t)\exp\left(-j2\pi k\frac{t}{N_c}\right) \\ S_{i,u_i}(k)=\dfrac{1}{\sqrt{N_c}}\sqrt{P_{u_i}\delta_{i,u_i,m}^{-\alpha}}\displaystyle\sum_{t=0}^{N_c-1}s_{i,u_i}(t)\exp\left(-j2\pi k\frac{t}{N_c}\right) \\ H_{u,m}(k)=\displaystyle\sum_{t=0}^{N_c-1}h_{u,m}\exp\left(-j2\pi k\frac{t}{N_c}\right) \\ H_{u_i,m}(k)=\displaystyle\sum_{t=0}^{N_c-1}h_{u_i,m}\exp\left(-j2\pi k\frac{t}{N_c}\right) \\ N_m(k)=\dfrac{1}{\sqrt{N_c}}\displaystyle\sum_{t=0}^{N_c-1}n_m(t)\exp\left(-j2\pi k\frac{t}{N_c}\right). \end{cases} \tag{4}$$

The first term in (3) is the desired signal, the second term is the MUI, the third term is the CCI, and the last term is the noise component.

The received signals $\{R_m(k); m=0\sim N_r\}$ are then expressed in a matrix form as

$$\mathbf{R}(k)=\mathbf{H}_0(k)S_0(k)+\sum_{u=1}^{U-1}\mathbf{H}_u(k)S_u(k)+\sum_{i=1}^{B}\sum_{u_i=0}^{U_i-1}\mathbf{H}_{i,u_i}(k)S_{i,u_i}(k)+\mathbf{N}(k), \tag{5}$$

where $\boldsymbol{R}(k)=[R_0(k), R_1(k)\cdots R_{N_r-1}(k)]^T$, $\boldsymbol{H}_u(k)=[H_{u,0}(k)\ \ H_{u,1}(k)\ \ \cdots\ \ H_{u,N_r-1}(k)]^T$, and $\mathbf{N}(k)=[N_0(k)\ \ N_1(k)\ \ \cdots\ \ N_{N_r-1}(k)]^T$ with $[\cdot]^T$ representing the transpose operation.

3. SC-FDAAA

The structure of the SC-FDAAA transceiver in both CAN system and DAN system can be generalized and shown in Fig. 3. Binary data sequence is modulated and divided into a sequence of blocks of N_c data symbols. The last N_g symbols in each block are copied and inserted as CP into the guard interval (GI) and placed at the beginning of each block. The received signal is transformed by an N_c -point fast Fourier transform (FFT) into the frequency domain signal and SC-FDAAA weight control is then performed on each frequency as

$$\widetilde{R}(k) = \mathbf{W}^{T}(k)\mathbf{R}(k) \tag{6}$$

where

$$\mathbf{W}(k) = \left[W_0(k), \cdots, W_{N_r-1}(k)\right]^T. \tag{7}$$

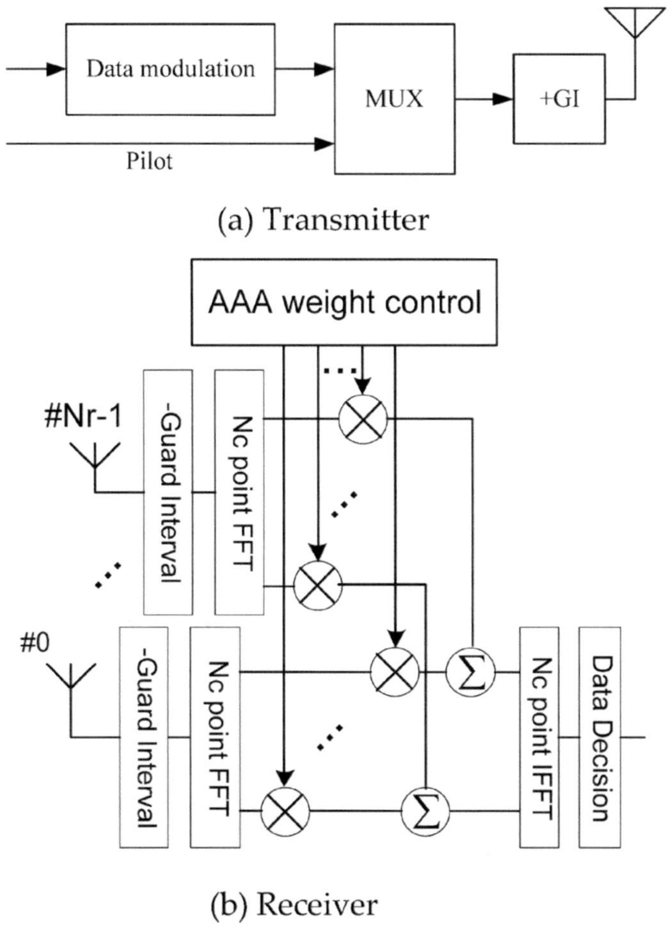

Figure 3. SC-FDAAA transceiver structure.

The SC-FDAAA weight that minimizes the mean squared error (MSE) between $\widetilde{R}(k)$ and the reference signal $S_0(k)$ (the pilot signal will be used as the reference signal) is given by [10, 11]

$$\mathbf{W}(k) = \mathbf{C}_{rr}^{-1}(k)\mathbf{C}_{rd}(k), \tag{8}$$

http://dx.doi.org/10.5772/57132

where $C_{rr}(k)=E\{\boldsymbol{R}^{*}(k)\boldsymbol{R}(k)\}$ is the correlation matrix of the received signal and $C_{rd}(k)=E\{\boldsymbol{R}^{*}(k)S_0(k)\}$ is the cross-correlation vector between the received signal and the reference signal, and * denotes complex conjugate operation.

$$\begin{aligned}\mathbf{C}_{rr}(k) &= E\left\{\mathbf{R}^{*}(k)\mathbf{R}(k)\right\} \\ &= \mathbf{A}_0^{*}(k)\mathbf{A}_0(k)+\sum_{u=1}^{U-1}\mathbf{A}_u^{*}(k)\mathbf{A}_u(k)+\sum_{i=1}^{B}\sum_{u_i=0}^{U_i-1}\mathbf{A}_{i,u_i}^{*}(k)\mathbf{A}_{i,u_i}(k)+N_0\mathbf{I} \\ &= \mathbf{A}_0^{*}(k)\mathbf{A}_0(k)+\mathbf{N}'(k),\end{aligned} \tag{9}$$

and

$$\mathbf{C}_{rd}(k)=E\left\{\mathbf{R}^{*}(k)S_0(k)\right\}=\mathbf{A}_0(k)S_0(k) \tag{10}$$

where $A_0(k)=\boldsymbol{H}_0(k)S_0(k)$ [12], $\boldsymbol{N}'=\sum_{u=1}^{U-1}\boldsymbol{A}_u^{*}(k)\boldsymbol{A}_u(k)+\sum_{i=1}^{B}\sum_{u_i=0}^{U_i-1}\boldsymbol{A}_{i,u_i}^{*}(k)\boldsymbol{A}_{i,u_i}(k)+N_0\boldsymbol{I}$ is used to represent the interference plus noise.

In the next, time domain signal block estimate is then obtained by an N_c - point IFFT for data decision as

$$\hat{d}(t)=\frac{1}{\sqrt{N_c}}\sum_{k=0}^{N_c-1}\tilde{R}(k)\exp\left(j2\pi k\frac{t}{N_c}\right). \tag{11}$$

4. Post SC-FDAAA SINR

The post SC-FDAAA SINR on the kth frequency can be evaluated by [13]

$$\Gamma(k)=\frac{\mathbf{W}^{H}(k)\mathbf{R}_s(k)\mathbf{W}(k)}{\mathbf{W}^{H}(k)\mathbf{R}_{N'}(k)\mathbf{W}(k)}, \tag{12}$$

where $\boldsymbol{R}_s(k)$ and $\boldsymbol{R}_{N'}(k)$ are the auto-correlation matrix of the received desired signal and the interference plus noise, respectively.

Property: if a matrix $\mathbf{Z}$ can be written as $\mathbf{Z}=\boldsymbol{T}^{-1}+\boldsymbol{P}\boldsymbol{Q}^{-1}\boldsymbol{P}^{*}$, then the inverse matrix of $\mathbf{Z}$ can be obtained by [14]

$$\mathbf{Z}^{-1} = \mathbf{T} - \mathbf{TP}\left(\mathbf{Q} + \mathbf{P}^{*}\mathbf{TP}\right)^{-1}\mathbf{P}^{*}\mathbf{T}. \tag{13}$$

Let $Z = C_{rr}(k)$, $T = R_{NI}^{-1}(k)$, $P = A_0^{*}(k)$ where $A_0(k) = H_0(k)S_0(k)$ and $Q = I$, then the inverse matrix $C_{rr}^{-1}(k)$ can be calculated by submitting Z , T , P and I into

$$\begin{aligned}\mathbf{C}_{rr}^{-1} &= \mathbf{R}_{N'}^{-1}(k) - \mathbf{R}_{N'}^{-1}(k)\mathbf{A}_0^{*}(k)\left[\mathbf{I} + \mathbf{A}_0(k)\mathbf{R}_{N'}^{-1}\mathbf{A}_0^{*}(k)\right]^{-1}\mathbf{A}_0(k)\mathbf{R}_{N'}^{-1}(k)\\ &= \mathbf{R}_{N'}^{-1}(k)\left[\mathbf{I} - \frac{\mathbf{A}_0^{*}(k)\mathbf{A}_0(k)\mathbf{R}_{N'}^{-1}(k)}{\mathbf{I} + \mathbf{A}_0(k)\mathbf{R}_{N'}^{-1}\mathbf{A}_0^{*}(k)}\right]\\ &= \left[\frac{1}{1 + \mathbf{A}_0(k)\mathbf{R}_{N'}^{-1}\mathbf{A}_0^{*}(k)}\right]\mathbf{R}_{N'}^{-1}(k).\end{aligned} \tag{14}$$

The SC-FDAAA weight is then obtained by substituting (9) and (13) into (7), given by

$$\mathbf{W}(k) = \left[\frac{1}{1 + \mathbf{A}_0(k)\mathbf{R}_{N'}^{-1}\mathbf{A}_0^{*}(k)}\right]\mathbf{R}_{N'}^{-1}(k)\mathbf{A}_0(k)S_0(k). \tag{15}$$

Finally, the SINR after the weight control can be expressed, by substituting (14) into (9), as

$$\Gamma(k) = \mathbf{A}_0(k)\mathbf{R}_{N'}^{-1}(k)\mathbf{A}_0^{*}(k). \tag{16}$$

5. Simulation results

Performance of DAN SC-FDAAA system will be investigated in this section, as a reference for comparison, the performance of CAN SC-FDAAA will also be evaluated. Cellular structures using FRF =1, 3, 4, 7, 9 and 12 will be considered. The parameters used to generate the results are listed in Tab. I. No channel coding is used for simplicity, and we assume that the transmit signal to noise ratio (SNR) is 10dB. The distributed antennas are located in a cell as shown in Fig. 4. In this study, scheduling among the distributed antennas is not considered. The scheduling algorithm and more complicated situation remain as the topics of our future work.

BER performance is investigated at first. In order to find out behaviors of both average BER and outage BER, the cumulative distribution functions (C.D.Fs) of BER performance are calculated and shown in Figs. 5-8 where the FRF equals to 1, 3, 4 and 7, respectively. For the cases of FRF 9 and FRF 12, the results are similar to the case of FRF 7, therefore those results are not shown for brevity. Fig. 5 shows a comparison between the C.D.Fs of BER in DAN system

http://dx.doi.org/10.5772/57132

and CAN system when FRF=1. The x axis is the BER abscissa and y axis is the probability that BER<abscissa. It can be observed that DAN SC-FDAAA outperforms CAN SC-FDAAA by having better BER performance. It can also be observed that when the number of users increases, the BER performance of SC-FDAAA will degrade in both DAN and CAN systems, which can be intuitionally expected due to the reduction of degree of freedom. From the results shown in Figs. 6-8, it can be further observed that when FRF increases, the C.D.F. curves of BER performance "shift" right-side, which means that the BER performance improves due to the reduction of CCI power. In addition, DAN SC-FDAAA always achieves better BER performance than CAN SC-FDAAA no matter how FRF varies. The results of BER performance have shown that the distributed nature of DAN system can significantly improve the BER performance of SC-FDAAA over CAN system.

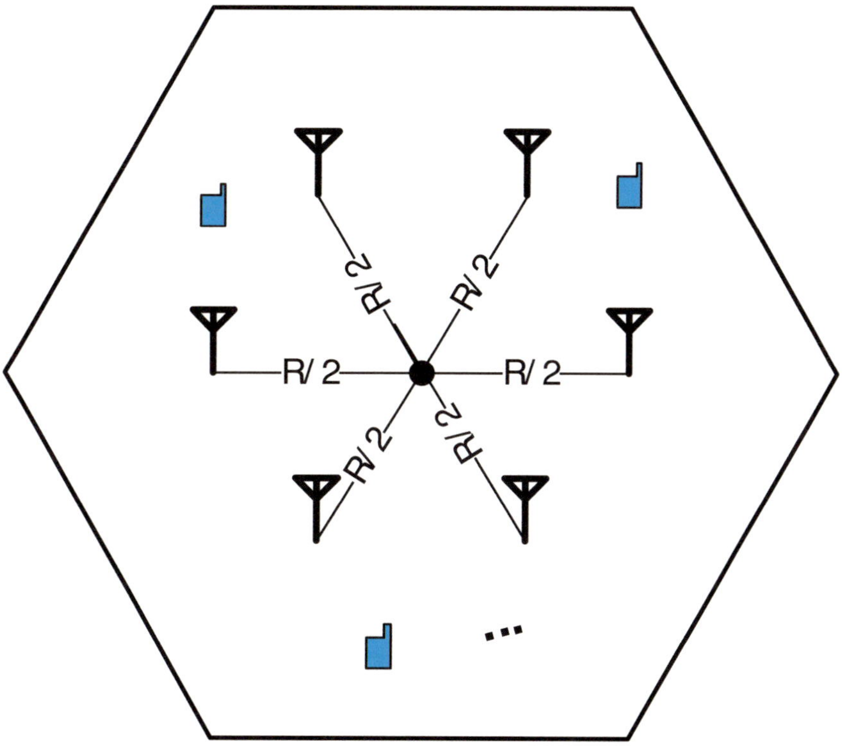

Figure 4. Antenna distribution in DAN system.

Modulation		QPSK
Channel	Channel Model	Frequency selective block Rayleigh fading
	Number of paths	$L = 16$
	Power delay profile	Uniform
	Path loss	$\alpha = 3.5$
SNR		10dB
Number of co-channel cells		$B = 6$
Number of antennas of mobile user		1
Number of users per cell		$U = 2$~6
User location distribution		Random
Number of antennas		$N_r = 6$
FFT (IFFT) points		$N_c = 256$

Table 1. Simulation Parameter

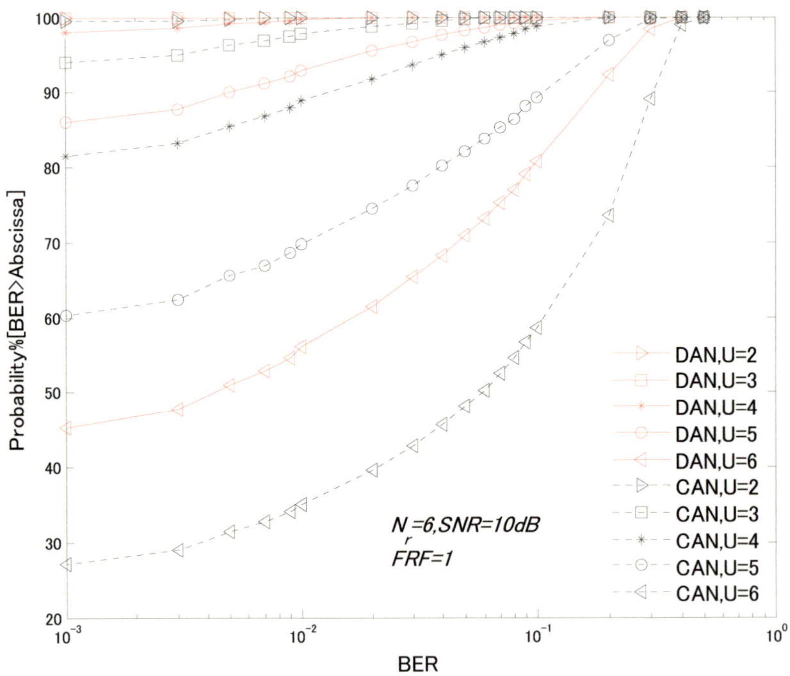

Figure 5. C.D.F. of BER performance, FRF=1.

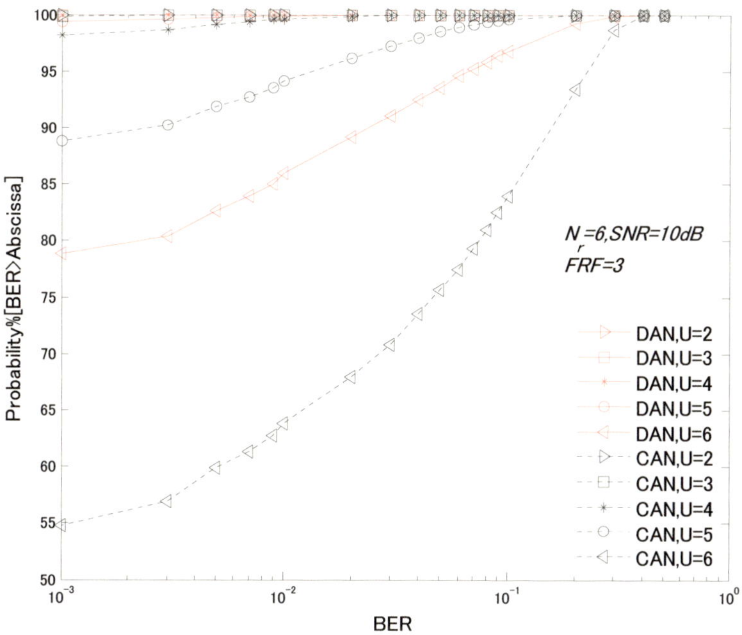

Figure 6. C.D.F. of BER performance, FRF=3.

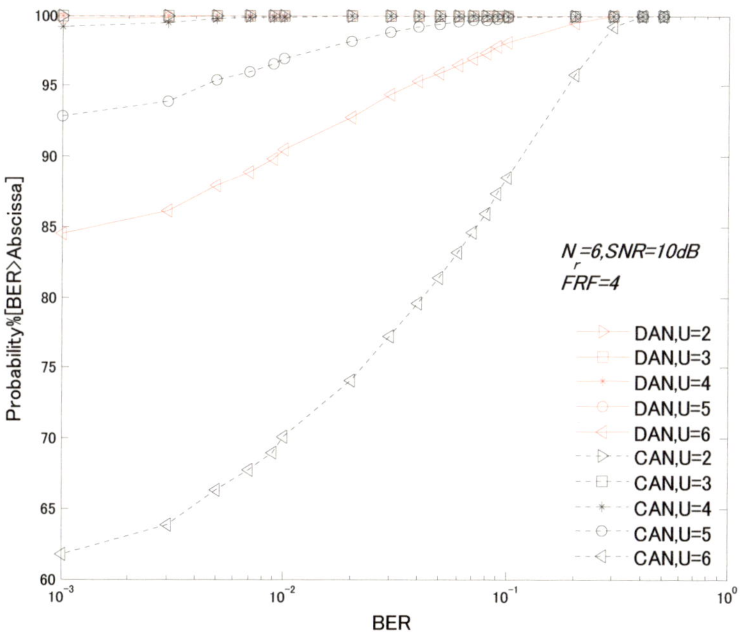

Figure 7. C.D.F. of BER performance, FRF=4.

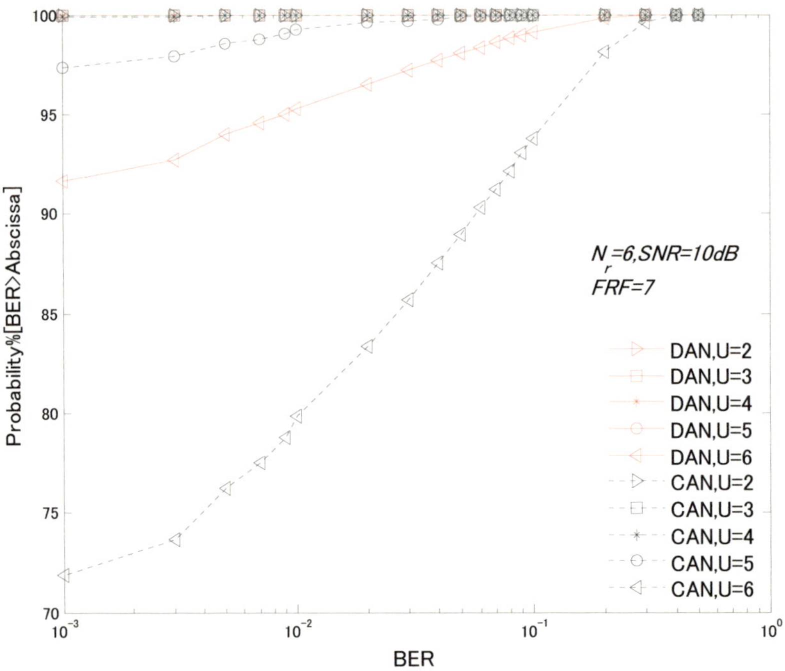

Figure 8. C.D.F. of BER performance, FRF=7.

System capacity given by bits/s/Hz can be calculated by Shannon capacity definition [15-16] using the SINR given in (15). However, the number of users that can be accommodated is a practical criterion to be considered as system capacity. Therefore, the following results will focus on the number of simultaneous users instead of the value given by bits/s/Hz. The average BER performance of DAN SC-FDAAA as a function of FRF is shown in Fig. 9. Since un-coded system is assumed, average BER=10^{-2} is used as a criterion to see how many users can be accommodated by using DAN SC-FDAAA. In the next, link capacity (maximum number of users/cell) and cellular link capacity (link capacity/FRF) of DAN SC-FDAAA are evaluated and the results are shown in Fig. 10 and Fig. 11.

Fig. 10 shows the link capacity of DAN SC-FDAAA and CAN SC-FDAAA. It is shown that 4 users can be accommodated by DAN SC-FDAAA when FRF 1 is used while 2 users can be accommodated by CAN SC-FDAAA, therefore, the link capacity can be doubled by using DAN SC-FDAAA. As the FRF increases, link capacities increase as the CCI power decreases. And when FRF is larger than 4, 6 users can be accommodated. Since N_r=6 and the AAA receiver can deal with up to N_r-1 interference, it can be concluded that the maximum number of users/cell of the DAN SC-FDAAA can approach its maximum value when FRF is larger than 4.

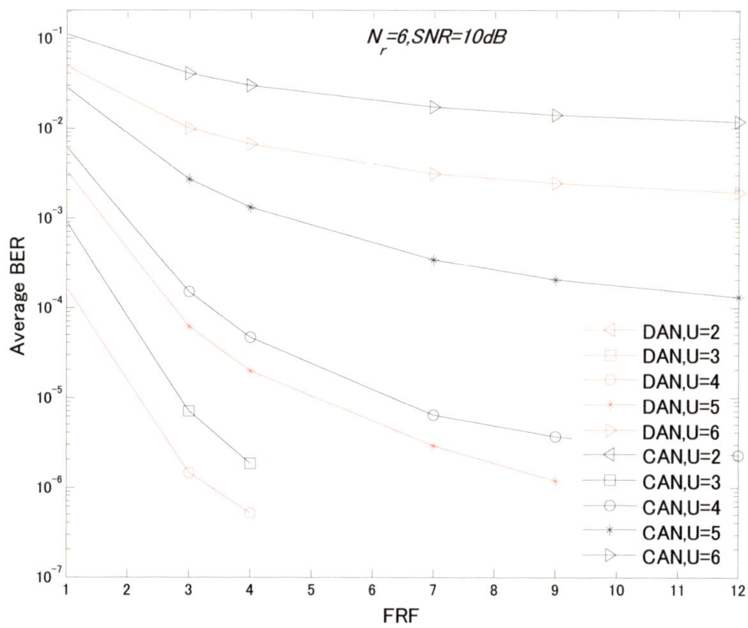

Figure 9. Average BER.

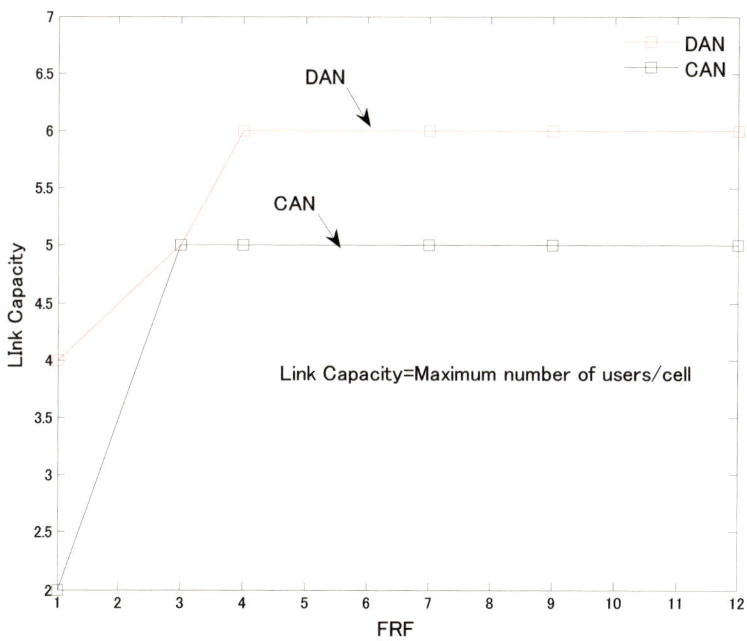

Figure 10. Link capacity.

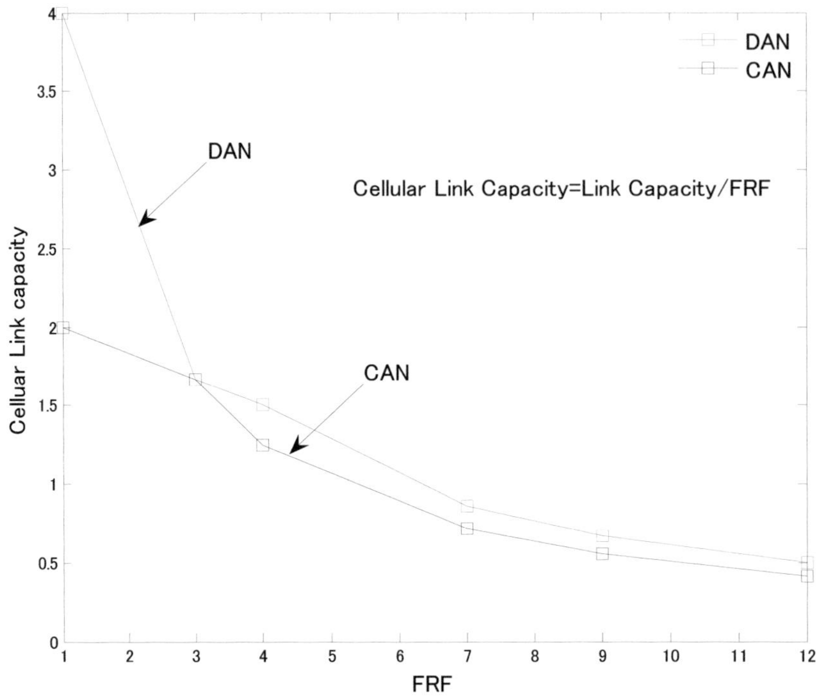

Figure 11. Cellular link capacity.

Larger FRF means more bandwidth will be consumed. In order to measure the spectrum efficiency, cellular link capacity is calculated and the results are shown in Fig. 11. It is shown that cellular link capacity of DAN SC-FDAAA can achieve its' maximum value when FRF =1 and decreases when FRF increases. Note that in our previous work on the cellular link capacity for conventional cellular system [17], it has been pointed out that the cellular link capacity can be maximized by using FRF=1 in the area near cell center and FRF = 3 in the area near the cell edge. Therefore, by using the DAN SC-FDAAA, a smaller FRF can be used and the spectrum efficiency can be greatly improved as a result. In addition, taking FRF=1 as an example, DAN SC-FDAAA achieves twice of the cellular link capacity as CAN SC-FDAAA does.

6. Conclusions

In this article, we presented the performance of DAN SC-FDAAA and also the performance of CAN SC-FDAAA as a reference. Both the BER distribution and capacity (link capacity and cellular link capacity) are given. The cellular link capacity can be maximized when FRF=1 by DAN SC-FDAAA. Therefore, the DAN SC-FDAAA can use the single frequency reuse (i.e., FRF=1) and improve the spectrum efficiency greatly. In addition, the performance of DAN SC-FDAAA can benefit from the distributed nature of DAN, and it can double the cellular link capacity by using the single frequency reuse when compared with CAN SC-FDAAA.

http://dx.doi.org/10.5772/57132

Author details

Wei Peng and Fumiyuki Adachi

*Address all correspondence to: peng@mobile.ecei.tohoku.ac.jp

Tohoku University, Japan

References

[1] J. G. Proakis, Digital Communications, fourth edition, New York: McGraw Hill, 2001.

[2] R. Price and P. E. Green, "A Communication Technique for Multipath Channels," Proc. IRE, vol. 46, pp. 555-570, March 1958.

[3] D. Falconer, S. L. Ariyavistakul, A. Benyamin-Seeyar and B. Edison, "Frequency Domain Equalization for Single-carrier Broadband Wireless Systems," IEEE Communication Magazine, vol. 40, pp. 58-66, April 2002.

[4] R. Van Nee and R. Prasad, OFDM for Wireless Multimedia Communications, Arctech House Publishers, UK, 1999.

[5] F. Adachi, K. Takeda, T. Obara, T. Yamamoto and H. Matsuda, "Recent advances in single-carrier frequency-domain equalization and distributed antenna network," IEEE ICICS 2009, pp.1-5, March 2009.

[6] K. Sivanesan and N. C. Beaulieu, "Outage and BER of MRC Diversity in Band-limited Micro-cellular Systems with CCI," IEEE Communications Letters, vol. 9, Issue. 3, pp. 207-209, March 2005.

[7] J. Zhang and J. Andrews, J, "Distributed Antenna Systems with Randomness", IEEE Transactions on Wireless Communications, vol. 7, no. 9, pp. 3636 - 3646,Sept. 2008.

[8] W. Peng and F. Adachi, "Frequency Domain Adaptive Antenna Array for Broadband Single-Carrier Uplink Transmission," IEICE Transactions on Communications, vol. E94-B, no. 7, pp. 2003-2012, July. 2011.

[9] W. Peng and F. Adachi, "Single-Carrier Frequency Domain Adaptive Antenna Array for Distributed Antenna Network," IEEE ICCS, pp. 1-5, Dec. 2010.

[10] Ahmed EI Zooghby, Smart Antenn Engineering, Arctech House Publisher, 2005.

[11] Simon Haykin, Adaptive Filter Theory, New York: Prentice Hall, 2002.

[12] J. H. Winters, "Signal Acquisition and Tracking with Adaptive Arrays in the Digital Mobile Radio System IS-36 with Flat Fading," IEEE Trans. Vehicular Technology, vol. 42, pp. 377–384, Nov. 1993.

[13] K. Hamdi, "On the Statistics of Signal-to-interference plus Noise Ratio in Wireless Communications," IEEE Transactions on Communications, vol. 57, no. 11, pp. 3199 - 3204, Nov. 2009.

[14] M. A. Woodbury, "Inverting Modified Matrices," New Jersey: Princeton, 1950.

[15] C.E. Shannon, "A Mathematical Theory of Communication I," Bell System Technical Journal, pp. 379–423, July 1948.

[16] C.E. Shannon, "A Mathematical Theory of Communication II," Bell System Technical Journal, pp. 623–656, October, 1948.

[17] W. Peng and F. Adachi, "Hybrid Frequency Reuse Scheme for Cellular MIMO Systems," IEICE Trans. Communications, vol. E92-B, pp. 1641-1648, May 2009.

Section 5

Multi-User MAC Schemes for High-Throughput IEEE 802.11n

Chapter 5

Multiuser MAC Schemes for High-Throughput IEEE 802.11n/ac WLANs

Elli Kartsakli, Nizar Zorba, Luis Alonso and Christos Verikoukis

Additional information is available at the end of the chapter

http://dx.doi.org/10.5772/57129

1. Introduction

In the last decade, the Wireless Local Area Network (WLAN) market has been experiencing an impressive growth that began with the broad acceptance of the IEEE 802.11 standard [1]. Given the widespread deployment of WLANs and the increasing requirements of multimedia applications, the need for high capacity and enhanced reliability has become imperative. Multiple-Input Multiple-Output (MIMO) technology and its single receiving antenna version, MISO (Multiple-Input Single-Output (MISO), promise a significant performance boost and have been incorporated in the emerging IEEE 802.11n standard [2].

Several multiple antenna transmission techniques such as spatial multiplexing and transmit beamforming are used to provide rapid and robust point-to-point wireless connectivity. On the other hand, due to the inherent diversity of the MIMO channel, it is possible to achieve simultaneous point-to-multipoint transmissions and serve multiple users at the same time, through the same frequency. The MIMO multiuser transmission concept, where multiple users are served through different data streams, can increase the overall system capacity when compared to single-user MIMO transmission, where all streams are dedicated to just one user [3].

Even though IEEE 802.11n has been designed with MIMO technology in mind, its main focus is on maximizing throughput in point-to-point transmissions, through spatial multiplexing and mechanisms such as frame aggregation. Neither the standard nor the majority of related work consider any Medium Access Control (MAC) mechanisms for multiuser scheduling, thus leaving a significant MIMO capability unexploited. As accurately pointed out in [4], there is a need for low-complexity multiuser transmission schemes, especially for downlink communications.

Recently, the IEEE 802.11ac task group has been working on an amendment of the 802.11 standard aiming to extend the total network throughput beyond the gigabit-per-second barrier. The new draft standard contemplates, among other things, multiuser transmissions in the downlink but leaves many open issues, especially on multiuser scheduling.

Motivated by this open line of work, this chapter is dedicated to the investigation of solutions for the incorporation of multiuser capabilities in IEEE 802.11n/ac WLAN systems by using Cross-Layer (CL) information, while maintaining backward compatibility with the standard. The main contribution is the design of a number of opportunistic channel-aware multiple antenna MAC schemes that handle multiuser downlink transmissions and explore the advantages that can be gained by exploiting multiuser diversity. Backward compatibility with the existing 802.11n standard and possible integration within the new 802.11ac draft standard are also key issue.

The remaining part of this chapter is divided into six sections. Section 2 provides an overview of the state of the art on multiuser MAC layer protocols, a review of the multiuser mechanism included in the 802.11ac draft standard and the additional challenges that arise from simultaneous transmissions. Section 3 discusses the problem statement and presents the Employed beamforming transmission technique. The description of the proposed multiuser MAC schemes is given in Section 4. Section 5 provides the performance evaluation of the proposed schemes and discusses the obtained trade-offs and finally, Section 6 closes the chapter with some general conclusions.

2. State of the art

Multiuser transmissions require the use of multiple antenna transmission techniques and advanced signal processing at the Physical (PHY) layer, as well as more complex MAC layer schemes. An overview of the most representative transmission techniques for smart antennas and MIMO systems is given in Section 2.1, whereas the available multiuser MAC schemes are presented in Section 2.2. Finally, a brief description of the multiuser capabilities of the IEEE 802.11ac draft standard and the relevant open issues are presented in Section 2.3.

2.1. PHY layer techniques for multiple antenna systems

In recent years, the technology of smart antennas has been widely investigated in an effort to increase the capacity of wireless networks. A smart antenna system combines multiple spatially distributed antenna elements with intelligent signal processing algorithms that adjust the antenna radiation pattern in order to achieve some desired objective. Smart antennas can be classified into three categories according to their level of intelligence [5]:

- The *switched beam antennas* have the lowest intelligence and can employ beamforming towards specific, predefined directions.
- The *dynamically phased antennas* can determine the direction of arrival of a received signal and steer a beam towards that direction to enhance reception.
- Finally, the *adaptive array antennas* can additionally adjust their radiation pattern to null out interference sources.

MIMO systems employ smart antenna technology with a high level of intelligence, aiming to improve transmission rates and enhance reliability and robustness. The term MIMO implies the availability of at least two antennas at each end of the communication link. When multiple antennas are employed only at the transmitter side the system is known

http://dx.doi.org/10.5772/57129

as Multiple-Input Single-Output (MISO), whereas Single-Input Multiple-Output (SIMO) systems imply a single transmitting and multiple receiving antennas.

There are several techniques for the exploitation of multiple antennas at the PHY layer, schematically illustrated in Figure 1. A brief description of each technique will be given next, but more detailed explanation can be found in [6] and [7].

The most conventional techniques are beamforming and interference suppression, shown in Figure 1a. By means of beamforming, the received signal strength of a point-to-point communication link is increased thus resulting to higher supported data rates and extended coverage range. Interference suppression is achieved by steering the nulls of the antenna radiation pattern towards specific directions. This technique can be employed to reduce the interference produced by the transmitter but also to limit the received interference by other systems. As a result the link reliability and the spectral efficiency of the system are enhanced.

Another very efficient technique is spatial diversity that effectively mitigates multi-path fading and therefore provides increased robustness against errors (Figure 1b). Depending on whether the multiple antenna elements are placed at the receiver (SIMO) or the transmitter (MISO), the spatial diversity schemes can be classified as receive and transmit diversity, respectively. In receive-diversity schemes, independently faded copies (due to different propagation paths) of the same signal arrive at each antenna element of the receiver and are appropriately combined or selected to enhance reception [8]. In transmit-diversity schemes the same signal is transmitted over multiple antennas after some processing has taken place to ensure that the received multiple copies of the signal will be successfully separated by the receiver [9][10][11]. Clearly, in MIMO systems, joint receive and transmit diversity schemes can be implemented.

A very powerful transmit-diversity technique that achieves both diversity and coding gain is the Space-Time Coding (STC) that involves signal coding over space (multiple antennas) and time (multiple symbol times). There are two main approaches to STC design, the Space-Time Trellis Coding (STTC) [12] and the Space-Time Block Coding (STBC) [13][14]. STTC provides considerable coding and diversity gains with the cost of high decoding complexity. On the other hand, STBC is less efficient since it mainly offers diversity gain (and minimal or zero coding gain) but has the significant property of using linear decoding at the receiver.

Another PHY layer technique is spatial multiplexing (Figure 1c), according to which multiple independent data streams are simultaneously transmitted in the same frequency spectrum using multiple antennas. The receiver manages to extract the data streams from the received signal by employing spatial processing techniques that exploit multi-path fading. As a result, the throughput performance is increased. A very popular and spectral efficient spatial multiplexing scheme is V-BLAST (Vertical- Bell Laboratories Layered Space Time) [15]. As far as point-to-multipoint links are concerned, the spatial multiplexing of signals known as SDMA allows multiple simultaneous transmissions in the same frequency, thus multiplying the capacity of the system [16].

Summarizing, the main PHY layer techniques that are available in multiple antenna systems are beam-forming, interference cancellation, spatial diversity and spatial multiplexing. These techniques can be used separately or in combination, to obtain the desired effect. Finally, it has been demonstrated that there is a fundamental trade-off between diversity gain and spatial multiplexing gain that reflects to a design decision in favor of increased reliability or throughput, respectively [17][18].

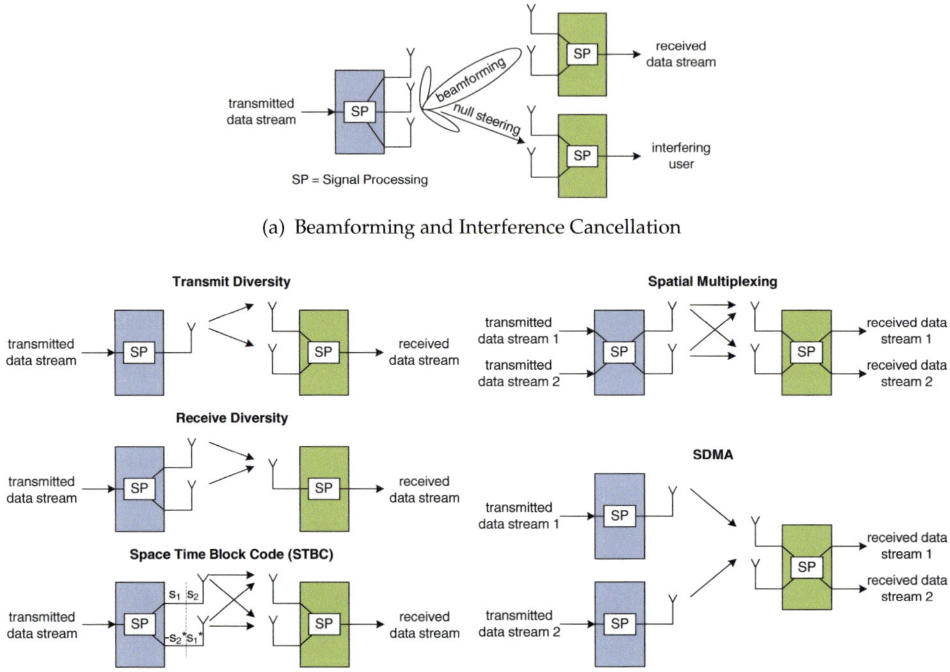

(a) Beamforming and Interference Cancellation

(b) Spatial Diversity Techniques

(c) Spatial Multiplexing Techniques

Figure 1. Multiple antenna transmission techniques

2.2. MAC protocols for multiuser transmissions

An overview of the most representative examples of multiuser scheduling and resource allocation can be found in [19]. The authors stress that selecting the best user subset for each transmission is the key to achieving multiuser diversity, but also point out several practical issues that must be taken care of, including the need for feedback acquisition on the channel state.

A significant number of contributions has been dedicated to the development of user selection and scheduling algorithms in the context of multiple antenna systems. An early work proposes the first-fit algorithm, a sub-optimum but less complex scheduling method that selects sets of packets that can be transmitted simultaneously [20]. However, one of the basic assumptions of this work is that the channel between the base station and the users is quasi-static and is considered known by the base-station, whereas scenarios with varying channel conditions are left for future consideration. In [21] the authors propose a SDMA/TDMA scheduler that assigns packets to time slots depending on their Quality of Service (QoS) requirements. Multiple packets can be spatially multiplexed in the same slot if they satisfy a Signal-to-Noise-and-Interference Ratio (SNIR) constraint. Again, this work mainly focuses on the scheduling policy and assumes that the spatial signature and QoS requirements for each packet are acquired during an initial admission phase.

Nevertheless, in realistic scenarios the channel condition cannot be considered known and a feedback mechanism must be established. Naturally, there is a trade-off between the potential performance enhancement when the channel is known and the reduced efficiency due to the introduced control overhead required for the feedback mechanism. One way to decrease feedback is by applying a threshold to exclude users with poor channel conditions from gaining access to the channel. This idea has been extensively studied in [22]. This work offers some guidelines for the threshold selection but it does not consider a specific multiple access scheme, nor the implementation of an actual feedback acquisition mechanism. In a different approach, binary feedback (1 or 0) is used by users to express whether they satisfy the threshold condition [23]. The idea is effective but assumes the presence of a dedicated low bit rate feedback channel, which is not the case in IEEE 802.11 based WLANs. Finally, another proposal combines the principle of splitting algorithms with threshold selection to determine the user with the best channel in less than three slots on average [24]. This work has been extended to provide detection of multiple users with good channel and needs on average 4.4 slots to find the best two users in the system [25].

Finally, there are some contributions that aim to include multiuser MAC schemes for IEEE 802.11 based systems. One example is the Multi-User Distributed Coordination Function (MU-DCF), presented in [26], that uses a four-way handshake that begins with a polling multiuser RTS frame. However there are several issues, mostly regarding the PHY layer implementation, that are not considered. A mathematical model for a downlink multiuser scheme for IEEE 802.11 is given in [27]. They show that performance can be improved by exploiting spatial multiplexing and conclude that there is still a need to design a modified MAC to support multiple transmissions and perform a good channel estimation mechanism.

2.3. Multiuser transmissions in the IEEE 802.11ac draft standard

In an effort to obtain WLAN throughputs beyond the gigabit per second barrier, a new draft standard, the IEEE 802.11 ac, is being developed, to extend the 802.11n capabilities in the 5 GHz band. The main target of the IEEE 802.11ac draft standard is to provide high aggregate throughput beyond 1 Gbps. The task group is currently in the process of developing the draft 7.0 version of the standard, with the final approval of the amendment expected towards the first quarter of 2014.

An important innovative feature of IEEE 802.11ac is the support of point-to-multipoint transmissions that are possible thanks to the multiuser capability of MIMO systems. In other words, a MU-MIMO capable device can transmit multiple packets simultaneously to multiple destinations. A maximum number of four users can be simultaneously supported and up to eight spatial streams can be employed for transmissions (with a maximum of four spatial streams per user).

The standard is not yet in its final form, but the most prevailing approach so far for the scheduling of multiple data frames is presented in [28]. The authors propose some modifications to the IEEE 802.11 backoff procedure and introduce new mode known as sharing of the transmission opportunity limit (TXOP). The main idea is that when a station gains access to the channel, it may be allowed to transmit simultaneously multiple packets that may belong to different access categories (i.e., traffic priorities), something that was not permitted in previous versions of IEEE 802.11. However, the exact rules of packet selection among the different access categories, and more generally, user and resource allocation issues and rate adaptation are not explicitly defined in the IEEE 802.11ac draft standard and pose interesting open issues.

3. System setup and multiuser physical layer

This section discusses the problem statement and describes the considered setup. In continuation, a brief description of the channel model is given and the multiuser transmission technique used at the PHY layer is explained. In general, the IEEE 802.11n MIMO specification with OFDM has been considered as the base for the PHY layer, with some modifications that will be explained in this section.

3.1. Problem statement and system setup

As indicated in the state of the art, it can generally be said that most contributions on multiuser transmission schemes focus on particular aspects of the problem and simplify the rest. Usually, when the focus is laid on the PHY layer transmission techniques, practical mechanisms for the channel access and the feedback acquisition are not considered, whereas multiuser MAC schemes often fail to consider PHY layer implementation issues. For example, some schemes optimize resource allocation but ignore feedback mechanisms and others minimize the required feedback but assume a dedicated control channel and a less sophisticated scheduling policy.

This chapter will introduce a multiuser MAC mechanism that handles in a joint manner the processes of channel access, scheduling, channel estimation and feedback acquisition, in conjunction with a low-complexity beamforming technique at the PHY layer. The proposed schemes have been designed in the context of a downlink communication channel in an infrastructure WLAN in which multiple antennas are available at the transmitter side. Without loss of generality, a MISO scenario with single-antenna users has been considered, even though the presented analysis can be also applied to MIMO systems with multiple-antenna users.

The considered setup is illustrated in Figure 2. The proposed schemes can be considered as a downlink transmission phase, initiated by an Access Point (AP) equipped with n_t antennas ($n_t \geq 2$) in a system with N single-antenna users. By exploiting the MIMO/MISO spatial signal processing capabilities and employing an appropriate transmission technique, the AP can serve up to n_t users at the same frequency and time. Nevertheless, in order to extract multiuser diversity gain, the pool of served users should exceed the number of transmitting antennas (i.e., $N > n_t$). Transmitting multiple downlink packets simultaneously, however, is feasible only when there is no interference among the selected users, or in a more realistic case, when the interference is relatively low. Hence, the AP must have some knowledge of the channel to select the most appropriate set of users for each transmission. These issues must be handled by the MAC layer in a practical way, as it will be described in detail in the following sections.

3.2. MIMO/MISO channel

With the use of OFDM, the frequency selective MIMO/MISO channel is transformed into a number of frequency flat channels. In particular, a block-fading model is considered

http://dx.doi.org/10.5772/57129

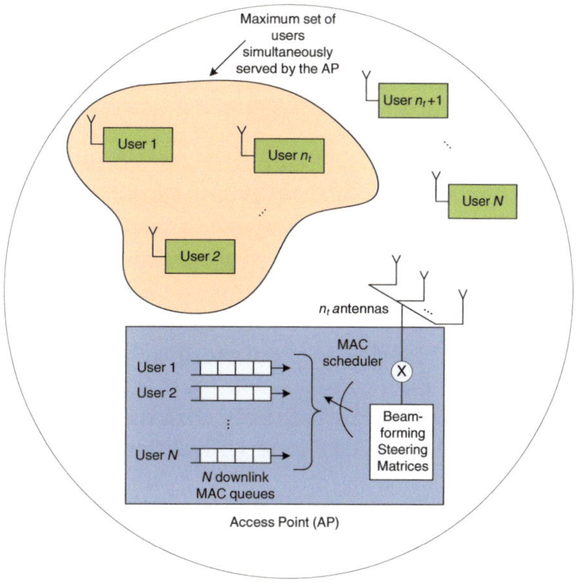

Figure 2. Scenario setup

for the channel which remains constant during the coherence time and changes between consecutive time intervals with independent and identically distributed complex Gaussian entries $\sim \mathcal{CN}(0, 1)$. This model represents the IEEE 802.11n channel model B in NLOS conditions [29], assuming that there are no time correlations among the different blocks and that the channel impulse response changes at a much slower rate than the transmitted baseband signal.

In the considered MISO downlink scenario, the channel between the AP that is equipped with n_t antennas and the ith single-antenna user (out of N total users with $N > n_t$) is described by a $1 \times n_t$ complex channel matrix $\mathbf{h}_i(t)$. Let $\mathbf{x}(t)$ be the $n_t \times 1$ vector with the transmitted signal to all the selected users in a particular transmission sequence and $y_i(t)$. Then, the received signal for the ith user can be expressed as

$$y_i(t) = \mathbf{h}_i(t)\mathbf{x}(t) + z_i(t) \tag{1}$$

where $z_i(t)$ is an additive Gaussian complex noise component with zero mean and $E\{|z_i|^2\} = \sigma^2$ is the noise variance. The transmitted signal $\mathbf{x}(t)$ encloses the independent data symbols $s_i(t)$ to all the selected users with $E\{|s_i|^2\} = 1$. A total transmitted power constraint $P_t = 1$ is considered and for ease of notation, time index is dropped whenever possible.

3.3. Multibeam Opportunistic Beamforming (MOB)

Multibeam Opportunistic Beamforming (MOB) is a low-complexity transmission technique for multiple-antenna broadcast channels [30]. MOB requires the presence of multiple antennas at the transmitter side and one or more antennas at each receiving user, meaning that it can be applied to MISO or MIMO scenarios. Its goal is to exploit multiuser diversity by finding a set of orthogonal users that can be simultaneously served on orthogonal beams, while maintaining the interference low. The key advantage of this transmission scheme is that it only requires partial Channel State Information (CSI) at the transmitter side in terms of the user received SNIR, making it very suitable for multiuser downlink communications.

The main steps of MOB are illustrated in Figure 3. It should be mentioned that these steps describe the main concept behind the MOB scheme without entering into implementation details. These will be more thoroughly addressed in Section 4 where the description of the proposed multiuser MAC schemes will take place. At the beginning of each transmission sequence, the AP forms n_t random orthogonal beams, equal to the number of its transmitting antennas (plot (a)). The users measure the SNIR related to each beam, select the highest measured SNIR value to the AP (plot (b)). In turn, the AP selects the best user for each beam and initiates the downlink data transmission (plot (c)). The scheme presented in [30] involves the opportunistic transmission by the users with the highest instantaneous SNIR for each beam, although MOB can also be combined with different scheduling policies.

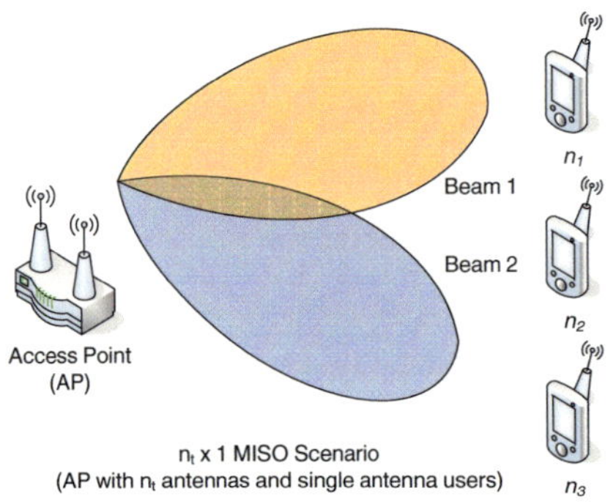

(a) MOB Step 1: The AP generates n_t random orthonormal beams

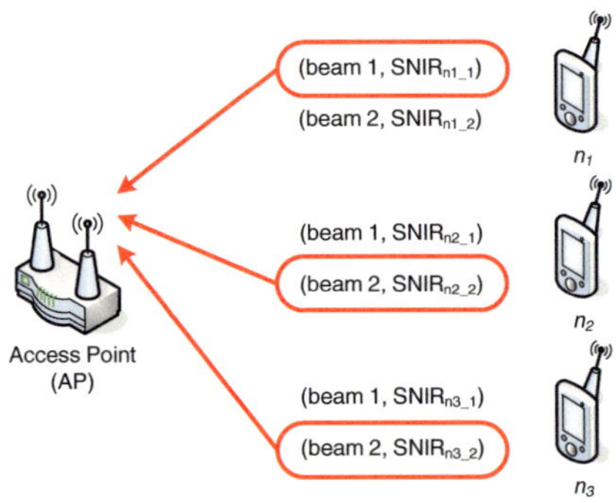

(b) MOB Step 2: Users measure the SNIR on each beam and feed back their best value

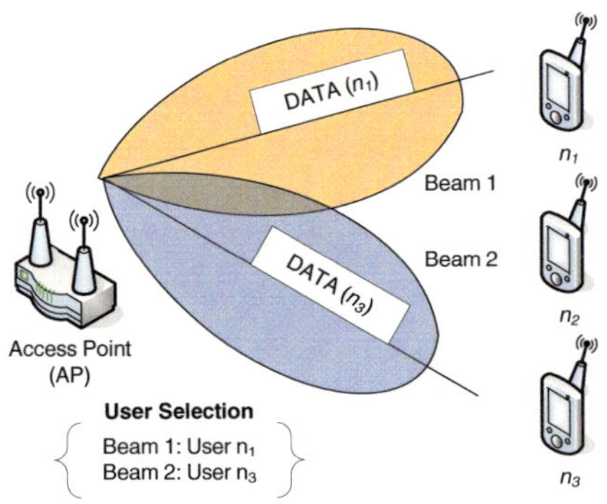

(c) MOB Step 3: The AP maps best users on beams and begins downlink transmission

Figure 3. Basic steps of MOB transmission technique

Through this low-complexity processing based on the instantaneous SNIR values, the MOB scheme achieves a high system sum rate by spatially multiplexing several users at the same time. In the best case where n_t users are selected for downlink transmission, the transmitted signal $\mathbf{x}$ can be expressed as

$$\mathbf{x} = \sqrt{\frac{1}{n_t}} \sum_{k=1}^{n_t} \mathbf{b}_k s_k \tag{2}$$

where s_k are the data symbols that correspond to the kth selected user, $\mathbf{b}_k$ is the assigned unit-power beam and the square root term is employed for total power constraint.

Although the beams are orthogonally generated, some of this orthogonality is lost in the propagation channel [30]. Consequently, some interference is generated by each beam on non-intended users. The SNIR formulation for the kth user that is served by the vth beam is

$$SNIR_{k,v} = \frac{\frac{1}{n_t} |\mathbf{h}_k \mathbf{b}_k|^2}{\sigma^2 + \sum_{u \neq v}^{n_t} \frac{1}{n_t} |\mathbf{h}_k \mathbf{b}_u|^2} \tag{3}$$

where a uniform power allocation is considered. The numerator is the received power from the desired beam, while the denominator represents the noise plus the interference power from the other beams.

As the number of users N grows, the AP can search for users in a larger pool, thus increasing the probability of finding a set of n_t users that do not interfere a lot among themselves [30]. Obviously, having $N \approx n_t$ results in an interference limited system, but for more practical values, such as $n_t = 2$ transmit antennas and $N \geq 10$ users, this scheme is efficient and has been shown to obtain higher performance with respect to single user opportunistic beamforming [31], [32].

The IEEE 802.11n PHY layer specification does not contemplate multiuser transmissions, even though it supports beamforming as a means to achieve higher data rates in point-to-point communications. Since the MOB scheme is practically a random beamforming transmission technique, it can be easily implemented within the standard without any further requirements in terms of hardware. The only necessary modification is to set accordingly the values of the beamforming steering matrices defined in the standard in order to form the random orthonormal beams.

4. Multiuser MAC schemes

The MOB technique is a low-complexity transmission scheme that can be easily implemented at the PHY layer to provide multiuser downlink communications. In a practical system, however, the beamforming scheme must be accompanied by a set of MAC layer functions to collect the necessary feedback information and handle the additional challenges that stem from simultaneous multiuser transmissions. This section will present three MAC layer schemes that modify the IEEE 802.11n MAC protocol to account for the demands and restrictions of the MOB technique. The required modifications are easy to implement within the IEEE 802.11n/ac standards and are backward compatible with the legacy single user transmission, in the sense that MOB and legacy users can coexist in the system.

Since the proposed MAC schemes aim to support the MOB transmission technique, they provide a common set of functions, graphically shown in Figure 4. These functions provide a practical MAC layer implementation to complement the three steps of the MOB scheme, namely the generation of the orthonormal beams, the acquisition of CSI feedback and the multiuser downlink transmission. In continuation, it is convenient to first present the common framework that applies to the three proposed schemes before proceeding with their detailed description that will focus on their differences in terms of complexity and efficiency.

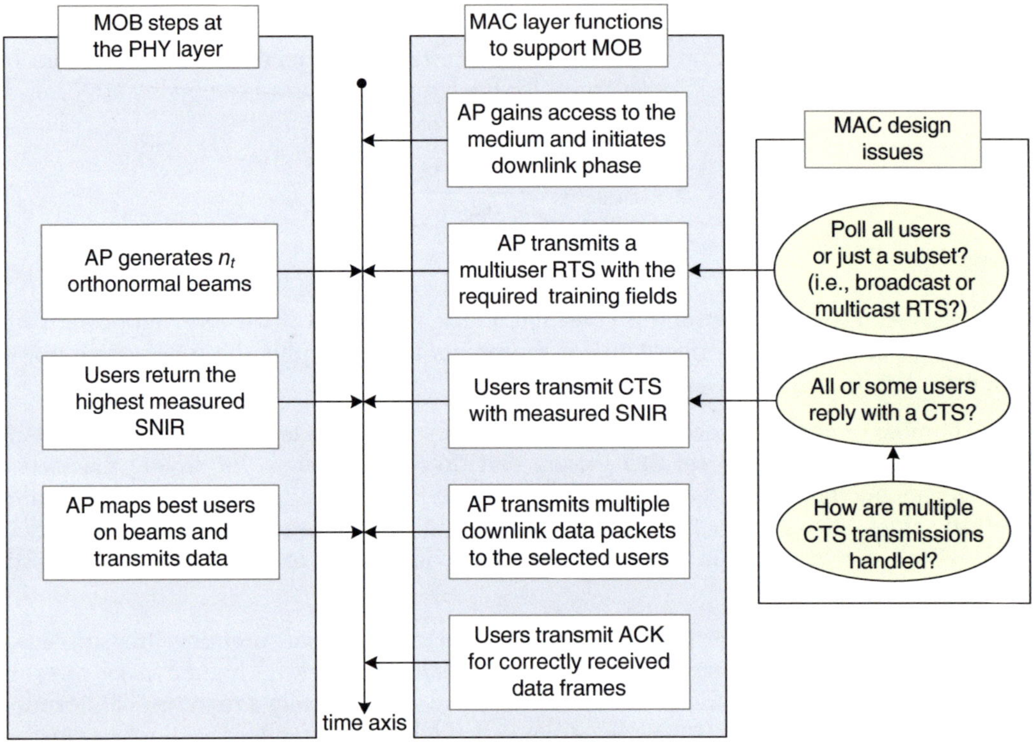

Figure 4. MAC layer functions to support the MOB transmission technique

As illustrated in Figure 4, the common functions provided by the MAC layer schemes are:

- *The initiation of the downlink phase.* The proposed multiuser schemes constitute a downlink phase that is always initiated by the AP, so for the sake of simplicity the backoff mechanism defined in the IEEE 802.11 specification is not employed in this study. Generally, in a scenario with both uplink and downlink transmissions, the AP would have to follow the backoff rules to gain access to the medium before initiating the downlink phase.
- *The generation of a multiuser RTS frame.* The initiation of the downlink phase is marked by the transmission of a modified RTS frame that basically serves two purposes:
  1. It is a call for participation in the downlink phase that may be addressed to a subset or to all the associated users (i.e., multicast or broadcast). The employed receiver address

http://dx.doi.org/10.5772/57129

included in the RTS is a point of differentiation between the proposed schemes and will be discussed later in this section.

2. It acts as a sounding frame that will enable the receiving users to measure the SNIR on each of the n_t generated beams. For this reason, the PHY layer preamble of the RTS contains a number of HT-LTFs (High-Throughput Long Training Fields), as defined in IEEE 802.11n standard. Apart from the training fields, the main body of the RTS frame is transmitted conventionally (i.e., on a single beam).

The structure of the modified RTS frame is shown in Figure 5. The length of the PHY layer preamble of the RTS frame is determined by the number n_t of spatial streams (i.e., orthonormal beams and subsequently antennas). For a single-antenna transmission, a PHY layer header of 28 μs is introduced, whereas for every additional spatial stream an extra HT-LTF of 4 μs is required. The description of the PHY header fields is given in Table 1 and more details can be found in the IEEE 802.11n specification [2].[1] The length of the MAC header mainly depends on the receiver address field. When a single receiver address is employed, the MAC header has a length of 20 bytes. Nevertheless, some of the proposed MAC schemes include multiple destinations in this address field, as it will be further clarified later.

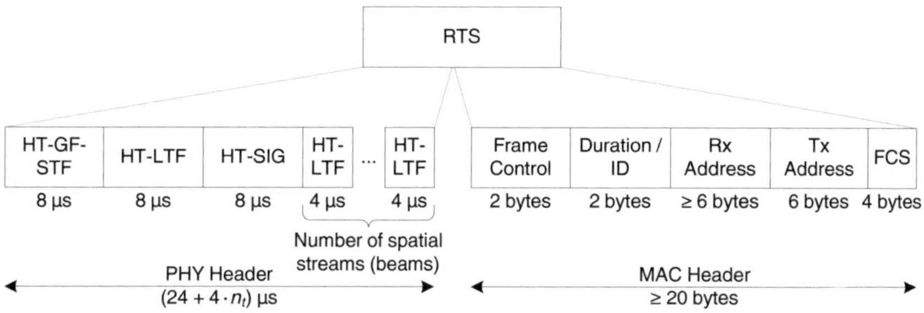

Figure 5. Structure of the modified RTS frame

Element	Description
HT-GF-STF	High-Throughput (HT) Greenfield Short Training Field
HT-LTF1	First HT Long Training Field
HT-SIG	HT SIGNAL Field
HT-LTF	HT Long Training Field

Table 1. Elements of the PHY layer header for the Multiuser MAC schemes

- *The transmission of a CTS frame by the downlink users.* Once the users receive the RTS frame and estimate their channel quality, they reply with a CTS frame that, unless otherwise stated, contains the best measured SNIR value and an integer identifier that corresponds to the respective beam. The structure of the modified CTS frame is shown in Figure 6. Assuming single-antenna users, a 28 μs PHY layer preamble is required,

[1] The PHY layer header structure presented in this section has been based on the IEEE 802.11n greenfield operation mode meant for IEEE 802.11n-only compatible stations. If compatibility with legacy devices is desired, the PHY layer headers should be modified accordingly, as indicated in Clause 20 of the IEEE 802.11n specification [2].

whereas the MAC header complies with the IEEE 802.11n specification, with the addition of an extra 1-byte field that contains the CSI information (i.e., the SNIR and the beam identifier).[2] Two design issues arise at this point. The first is whether a CTS should be transmitted by every polled user, or a limit should be posed to the number of CTS replies, for example by filtering out users with very bad channel conditions. The second issue concerns the transmission order of the CTS frames by the multiple users which can be either deterministic, thus collision-free, or random (probabilistic) that will likely result to collisions among simultaneously transmitted CTS. These two issues will be handled in different ways by the proposed MAC schemes, as it will be discussed later.

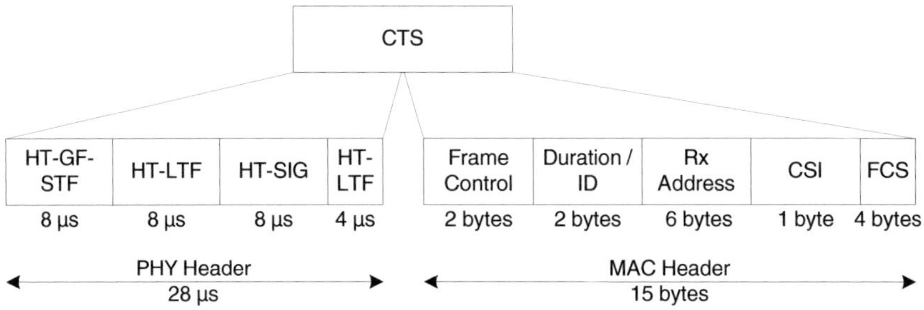

Figure 6. Structure of the modified CTS frame, including CSI feedback

- *The transmission of multiuser data frames by the AP.* Once the AP collects the feedback information included in the CTS frames it assigns the user with the highest measured SNIR on each beam (at most one user per beam) and transmits a maximum of n_t data packets simultaneously. The data packets employ the channel over the same time, frequency and code but are transmitted over different beams. This can be supported by the IEEE 802.11n standard, by exploiting the multiplexing capabilities of multi-antenna systems. This is actually an important shift from current systems where the simultaneous transmission of multiple packets in the same medium leads to collision and packet loss. Link adaptation is also employed and the transmission rate on each beam is determined by the measured SNIR.
- *The transmission of ACK frames.* The users signal the correct reception of a data frame by transmitting an ACK. In the proposed schemes, the multiple (up to n_t) ACK frames are transmitted sequentially, following the mapping of the users onto the beams.

In the remaining part of this section, the three proposed MAC layer schemes will be described in detail.

4.1. Mu-Basic scheme

The first and simplest scheme is called *Mu-Basic* and is a straightforward adaptation of the IEEE 802.11 mechanism to support downlink multiuser transmission. This scheme is based on the principle that at most n_t users can be served simultaneously by an AP equipped with n_t transmitting antennas that generate an equal number of orthogonal beams. Hence, in the

[2] In this work, it has been assumed that a SNIR quantization scheme has been employed so that the CSI field can be sufficiently represented by 1 byte.

http://dx.doi.org/10.5772/57129

beginning of the transmission sequence, the AP randomly selects n_t users from the downlink message buffer and transmits a multidestination RTS frame that includes the respective n_t receiver addresses, as illustrated in Figure 7.

The figure focuses on the receiver address field, since the remaining part of the RTS frame follows the structure indicated previously (Figure 5). The order in which the addresses are listed serves two purposes. First, it indicates the order in which CTS frames are to be sent in order to avoid collisions. Second, the address list is used to implicitly map the polled users to the beams. The users that receive the RTS frame check whether their address is in the list and wait for a predefined time before sending a CTS, which includes the SNIR measurement that corresponds to the assigned beam.[3] Note that in this case, the users do not reply with the best SNIR value since the beam assignment is predefined by the AP.

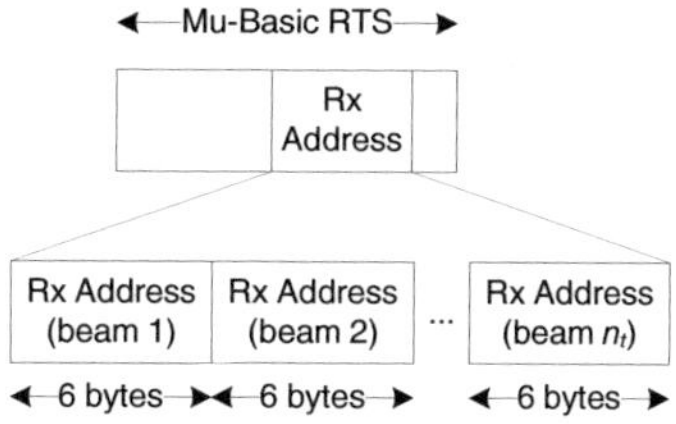

Figure 7. The modified RTS frame for the Mu-Basic scheme

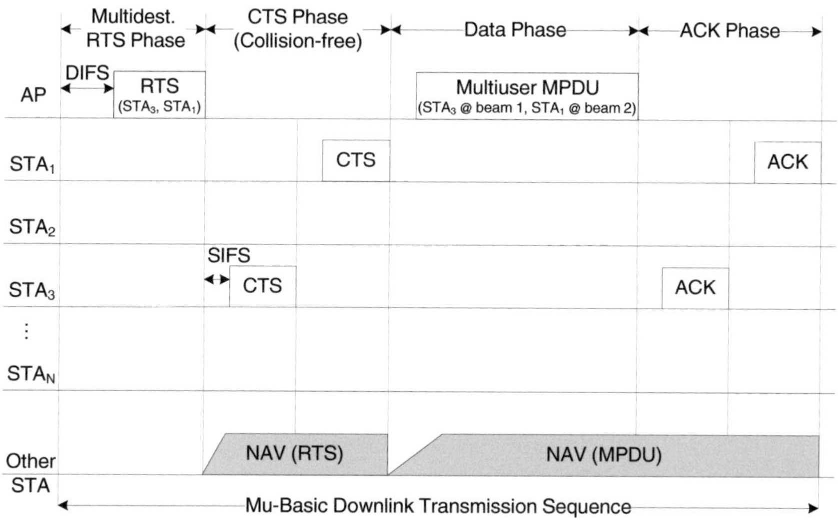

Figure 8. Transmission sequence example for the Mu-Basic scheme

The AP proceeds to the simultaneous transmission of the n_t data packets after selecting the transmission rate for each beam, according to the corresponding SNIR measurement that indicates the link quality. The users acknowledge the data reception by sequentially sending

[3] Since each CTS slot is of a fixed duration (i.e., a SIFS time and the time required for the transmission of the 15 byte CTS with the minimum available transmission rate) and assuming negligible propagation delays, each user can determine when to initiate the CTS transmission.

an ACK frame. An example of the transmission sequence according to the Mu-Basic scheme is given in Figure 8. In this example, there are $n_t = 2$ antennas at the AP, so two users are randomly selected for transmission (STA_3 and STA_1).

To avoid collisions by users that do not participate in the process, the IEEE 802.11 NAV mechanism can be employed. For this reason, the time from the transmission of the RTS until the end of the CTS phase is marked in the duration field of the RTS frame (Figure 5). The remaining time of the frame sequence, from the end of the CTS phase until the transmission of the last ACK, is indicated in the respective duration field of the data packet MAC header. Hence, non-participating users can set their NAV timer upon the RTS reception and can later update it when the header of a data packet is decoded.

Mu-Basic is easy to implement since it is a simple polling scheme initiated by the AP. Its performance will serve as a benchmark for the evaluation of the two more advanced multiuser schemes that will be presented next. In the considered case the destination users are randomly selected, however different criteria could also be applied to prioritize users with specific demands (e.g., with delay sensitive applications). Mu-Basic requires some additional overhead in the RTS frame as multiple receiver addresses must be included, but has the shortest possible CTS phase, since the number of received CTS frames is equal to the n_t served users (it would not make sense to receive feedback from less than n_t users if all the parallel streams were to be employed). On the other hand, multiuser diversity is not exploited since the users are scheduled without any consideration of their channel quality. Thus, the user selection and the beam assignment processes are not optimally done. As a result, the interference among the scheduled set of users may be high, leading to transmissions at low data rates (i.e. interference controlled system).

4.2. Mu-Opportunistic scheme

In an effort to exploit multiuser diversity and transmit opportunistically to the best set of users, according to the principles of the MOB transmission technique, the *Mu-opportunistic* scheme has been proposed. This scheme provides a mechanism for the AP to acquire the CSI of all users before reaching a scheduling decision, in order to optimize user selection and beam allocation. To this end, in the beginning of the transmission sequence, the AP polls all users with available data for downlink transmission. For the sake of simplicity, it will be assumed that the system is under saturation and there is always downlink traffic for each of the N system users.[4] Hence, the AP transmits a multidestination RTS frame that includes the N receiver addresses of all the network, as illustrated in Figure 9.

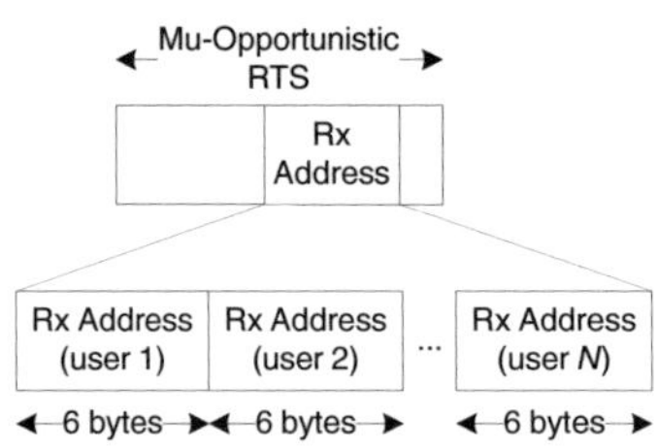

Figure 9. The modified RTS frame for the Mu-Opportunistic scheme

[4] The non saturation case will be examined later in this chapter.

http://dx.doi.org/10.5772/57129

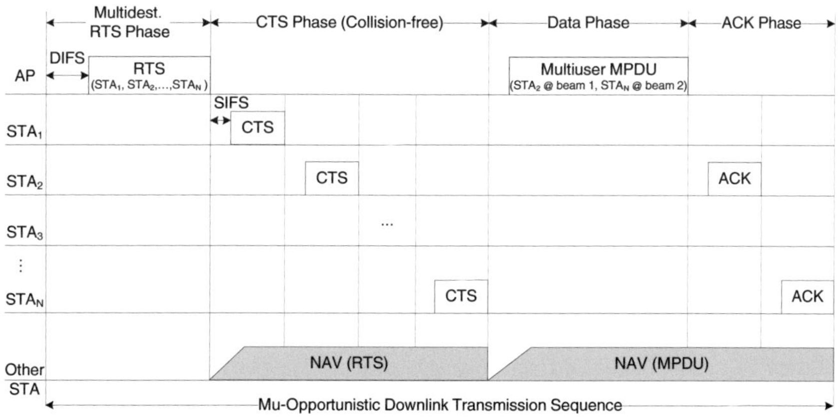

Figure 10. Transmission sequence example for the Mu-Opportunistic scheme

The users measure the SNIR on all the beams and include the maximum SNIR value in the CTS, along with an integer identifier of the beam that yielded that value. As before, CTS packets are transmitted in a collision-free manner, following the order of the address list in the RTS. After receiving all the feedback, the AP assigns each beam to the user with the highest SNIR and proceeds to the downlink data transmission. If a beam is not selected by any user then it is not used for transmission, even though this is not likely to happen very often for a large number of active users and a time-varying channel. Correct data reception is marked by the transmission of ACK frames that are sent sequentially, according to the beam allocation order (the user served on the first beam replies first, and so on). An example of the transmission sequence according to the Mu-Opportunistic scheme is given in Figure 10. In this example, there are $n_t = 2$ antennas at the AP and N users with available data. The AP receives N CTS frames and then selects the best set of users (STA_2 and STA_N, in the example) for the downlink data transmission.

The Mu-Opportunistic fully exploits multiuser diversity since it opportunistically schedules users with good channel conditions and with low mutual interference (i.e., users with high SNIR values measured on different beams). The weakness of this scheme is that it introduces significant overhead, mainly due to the long CTS phase, and the trade-off between overhead and efficiency becomes critical, especially as the number of users N grows.

4.3. Mu-Threshold scheme

The Multiuser Threshold-Selective algorithm (*Mu-Threshold)* is the third proposed multiuser MAC layer scheme. It maintains the opportunistic scheduling policy of selecting a set of users with high rates and low mutual interference but also aims to limit the additional control overhead. In order to achieve these objectives, it introduces two major changes with respect to the Mu-Opportunistic scheme:

- Instead of the deterministic, collision-free CTS transmissions, Mu-Threshold introduces a CTS contention phase during which users compete with each other within a predefined number of slots. Generally, even though collisions among CTS frames are likely to occur, the number of slots is smaller than the total number of users, thus reducing the length of the CTS phase.

- In order to reduce the CTS collision probability, the algorithm imposes a SNIR threshold so that only users with a relatively good channel are allowed to participate in the feedback process. Even though the idea of threshold application is not new, the novelty lies in the inclusion of this concept in a feasible MAC scheme for a multiuser MIMO scenario.

The frame exchange sequence of the Mu-Threshold scheme is initiated with the broadcast transmission of an RTS by the AP. The advantage of this configuration is that it calls all the users to participate in the CTS contention phase by employing a single 6-byte destination address instead of a long address list, as shown in Figure 11. Without doubt, this setup is meaningful under a saturation scenario in which the AP has always packets to transmit to all the associated users. This consideration is made to facilitate the evaluation of the full potential of the Mu-Threshold scheme, given that opportunistic downlink schemes are mostly needed under high-traffic conditions. In non-saturation conditions, the Mu-Threshold scheme could be applied with a minor modification. In this case, the AP would have to periodically set up multicast groups with the subset of active users (i.e., those who are waiting to receive downlink data) and use a multicast instead of a broadcast address.

After the RTS transmission, a CTS contention phase of m slots is initiated, with m being a system parameter subject to optimization. The slots have a predefined length, equal to a SIFS duration plus the time required for the transmission of the 15 byte CTS with the minimum available transmission rate. Depending on whether the maximum SNIR measured by a user is above or below the threshold, the user is either allowed to participate in this phase, or forced to remain silent until the beginning of a new frame sequence. Those allowed to participate select randomly a slot with equal probability and transmit a CTS containing the maximum measured SNIR and the corresponding beam identifier. Whenever multiple users select the same slot a collision occurs and the involved CTS frames are considered lost (the capture effect is not considered, even though it could increase the effectiveness of the proposed scheme). A slot can also remain empty if no user selects it for transmission.

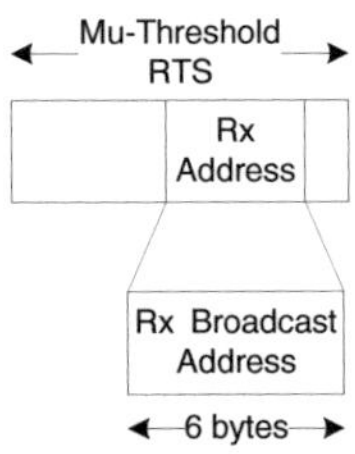

Figure 11. The modified RTS frame for the Mu-Threshold scheme

The next stage of the algorithm depends on the outcome of the contention phase. If no CTS has been correctly received (due to either collisions or lack of user participation because of the SNIR threshold value) no data is transmitted and a new contention phase is initiated.[5] User synchronization has been assumed, so that a collision in the mth slot only affects the involved CTS packets and does not have any effect on transmissions in the remaining slots of the contention phase. Thus, if at least one CTS is received, transmission of downlink data packets can take place. As before, the AP assigns the best user on each beam, based on

[5] Different policies could be implemented to avoid the presence of empty frames (e.g., transmission to a randomly selected user or to a user with a long waiting time using a basic rate) but will not be considered in this work.

http://dx.doi.org/10.5772/57129

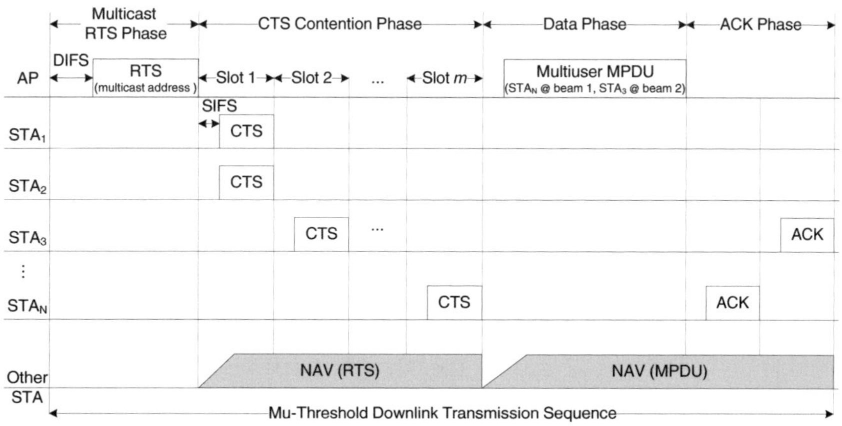

Figure 12. Transmission sequence example for the Mu-Threshold scheme

the feedback information collected by the received CTS frames and transmits a maximum of n_t data packets simultaneously. Note that, unlike the contention phase where collisions among CTS frames can occur, the transmission of data is collision-free. Finally, the users acknowledge the data reception by sequentially sending an ACK frame, following the order of the user mapping onto the beams.

An example of the transmission sequence according to the Mu-Threshold scheme is given in Figure 12. In this example, there are $n_t = 2$ antennas at the AP and N users with available data that compete in m CTS slots (with $m < N$ in general). Some users may select the same slot and collide (e.g., STA_1 and STA_2), others may transmit a CTS successfully (e.g. STA_3 and STA_N) and finally a number of users will refrain from this phase due to their unfavorable channel conditions.

An important decision is the selection of the SNIR threshold that serves two purposes: it reduces the number of contending users, thus decreasing the probability of CTS collisions, and it filters out those users with harsh channel condition, resulting to transmissions with higher data rates. Nevertheless, selecting a high threshold could cause adverse effects such as starvation if the majority of users experience low link quality. The threshold is determined by the AP and it is made known to the users during an initial association phase (alternatively, it could be included in the RTS packet, thus increasing its size by a few overhead bits). It is also possible to design a dynamic scheme that will adapt the threshold value at runtime depending on measured channel statistics.

The number of the CTS contention slots m is another important parameter that depends on the number of participating users which, in turn, is determined by the total number of users N, their channel condition and the selected threshold. An interesting observation is that, since the duration of each CTS slot is fixed, the duration field of the RTS packet (that indicates the length of the CTS phase) implicitly reveals the number of contention slots m. Therefore, the AP can let the users know the value of m without requiring an additional control field.

5. Performance evaluation

5.1. Simulation setup

This section will focus on the performance evaluation of the proposed multiuser schemes. Simulation results have been obtained with the help of a custom-made link layer simulation tool implemented in C++. Theoretical analysis of the proposed schemes has also been derived and more details can be found in [33].

The simulation setup considers an infrastructure downlink network that consists of an AP with $n_t = 2$ transmitting antennas and $N = 10$ single-antenna users (MISO scenario). An ideal Adaptive Modulation and Coding (AMC) that ensures error-free data transmission has been assumed at the PHY layer, given that the rate for each transmission is selected according to the link quality, as expressed by the SNIR.

A channel model that represents the IEEE 802.11n channel model B in Non-line-of-sight (NLOS) conditions has been considered [29]. As mentioned in Section 3.2, a block-fading model with independent and identically distributed complex Gaussian entries $\sim \mathcal{CN}(0,1)$ has been considered, with a noise variance of 0.1.[6] Each block corresponds to the duration of a frame sequence and no correlations have been assumed among the different blocks. This model has been employed to generate a SNIR matrix that represents the channel condition of each user on a frame-by-frame basis. The SNIR limits employed to determine the available transmission rate of each user are given in Table 2 [34].

Rate (Mbps)	SNIR (dB)
0 (no transmission)	$\leq$-8
6	-8 to 12.5
9	12.5 to 14
12	14 to 16.5
18	16.5 to 19
24	19 to 22.5
36	22.5 to 26
48	26 to 28
54	>28

Table 2. SNIR thresholds

Four different scenarios have been considered, characterized by four channel implementations (i.e., different SNIR matrices) denoted by Ch_A, Ch_B,Ch_C and Ch_D. The average link quality varies for each channel, with Ch_A corresponding to the most unfavorable conditions and Ch_D representing a channel with high quality links. For reference, the average user SNIR for channels Ch_A to Ch_D is 15dB, 17dB, 20dB and 25dB, respectively. According to Table 2, the average user rate for each scenario will be 12, 18, 24, and 36 Mbps, respectively. Since the channel realizations are random, the available rate for each user at every time

[6] Without loss of generality, a relatively low noise variance has been used. Higher values would lead to different numerical results but without affecting the behavior of the evaluated MAC schemes.

http://dx.doi.org/10.5772/57129

instance will oscillate around the mean value (with the same variance for all users), through the block fading channel defined in Section 3.2.

Parameter	Value
Number of antennas (AP)	$n_t = 2$
Number of antennas (Users)	$n_r = 1$
Downlink Users	$N = 10$
SIFS	16 μs
aSlotTime	9 μs
PHY Header (AP)	28 μs
PHY Header (Users)	32 μs
MAC Header	40 bytes
RTS (Mu-Basic)	$14 + 6 \cdot n_t$ bytes
RTS (Mu-Opportunistic)	$14 + 6 \cdot N$ bytes
RTS (Mu-Threshold)	20 bytes
CTS	15 bytes
DATA	2312 bytes
ACK	14 bytes
Bandwidth	20 MHz

Table 3. Simulation parameters

Saturated traffic conditions have been considered, with a constant flow of downlink traffic for all users always available at the buffers of the AP. The rationale behind this assumption has been to evaluate the maximum gain that can be extracted from downlink transmissions, which requires the system to operate under heavy traffic load. Unless otherwise stated, the size of the data packets has been fixed to 2312 bytes. All control frames are transmitted at the lowest rate (i.e., at 6 Mbps) to ensure correct reception. The IEEE 802.11n frame format has been adopted at the MAC layer, with the modifications proposed in Section 4 for each multiuser scheme. A summary of the simulation parameters is given in Table 3.

5.2. Performance comparison of the multiuser schemes

This section compares the performance achieved by the proposed multiuser MAC schemes, Mu-Basic, Mu-Opportunistic and Mu-Threshold. Figures 13 and 14 plot the throughput and mean total delay performance for the four channel models, $N = 10$ users and a packet size of $L = 2312$ bytes. The performance of the non-realistic Mu-Ideal scheme is also depicted, as a reference of the upper bound that corresponds to the considered scenarios. Mu-Ideal is an ideal opportunistic multiuser scheme in which the users with the highest SNIR values are scheduled on each beam. In other words, the same scheduling objective as in the Mu-Opportunistic scheme (Section 4.2) is targeted. The difference is that, in the Mu-Ideal scheme it has been assumed that the AP has a perfect knowledge of the channel condition and can select the best set of users without any additional overhead. Clearly, this scheme is not practical, since some mechanism for the CSI acquisition must be available at the AP.

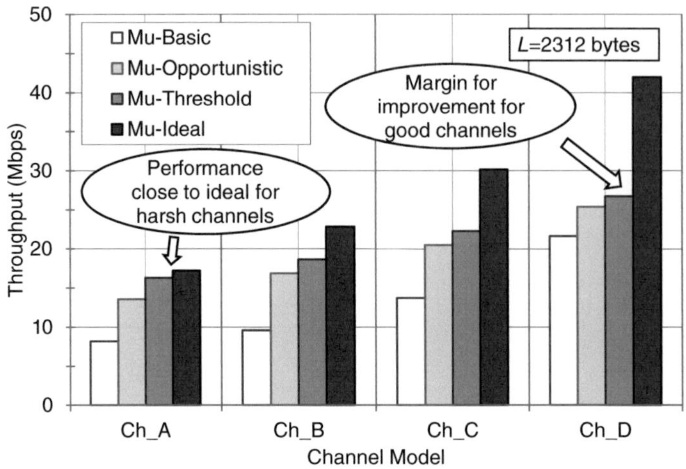

Figure 13. Throughput performance comparison

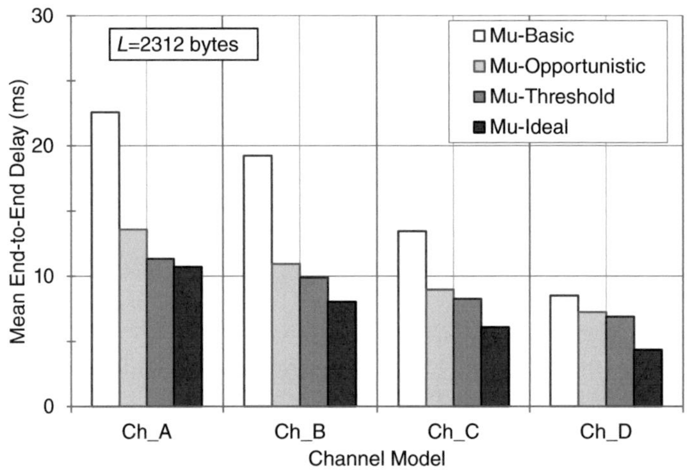

Figure 14. End-to-End delay performance comparison

The presented results for the Mu-Threshold have been obtained by considering the best combination of threshold and CTS slot number values. These optimum parameters are summarized in Table 4. In general, the channel statistics influence heavily the Mu-Threshold performance and the optimization of the algorithm is not straightforward since different objectives must be met to maximize performance in diverse scenarios. This can be better understood by examining the percentage of empty frames, given in the last column of the table. In the case of Ch_D, this percentage is low, meaning that the majority of frames feature single or double data transmissions. On the other hand, for harsh channels the minislot-threshold combination that maximizes throughput may result to a higher number of empty frames (even up to 50% for Ch_A), thus revealing that it is more efficient, as far as throughput is concerned, to transmit fewer packets but with a higher rate that to transmit in every transmit sequence with lower rates.

Channel	**Threshold** r_γ	**Slots** m	**Empty Frames** %
Ch_A	24	2	49.9
Ch_B	24	3	29.2
Ch_C	36	3	30.7
Ch_D	48	3	23.9

Table 4. Best configuration for Mu-Threshold scheme

The performance of the two opportunistic schemes, Mu-Opportunistic and Mu-Threshold, is bound between Mu-Basic and Mu-Ideal schemes. To illustrate this point, two performance statistics have been calculated in Table 5. The first metric reflects the throughput gain of the two schemes with respect to the Mu-Basic algorithm. It can be observed that both schemes improve performance under all the considered channel models by scheduling users with high available transmission rates. However, the exact value of the achieved gain depends on the channel quality. For harsh channels, the improvement is more pronounced. In the case of Ch_A, for instance, a gain of approximately 66 % and 99 % is obtained by Mu-Opportunistic and Mu-Threshold, respectively. On the other hand, when the channel quality is good, as in Ch_D, the need for opportunistic scheduling is less critical. Nevertheless, even in that case, an enhancement of more than 20 % can be achieved.

	Throughput gain (%) with respect to Mu-Basic		**Improvement margin (%) with respect to Mu-Ideal**	
Channel	*Mu-Opport.*	*Mu-Thres.*	*Mu-Opport.*	*Mu-Thres.*
Ch_A	65.84	99.21	26.80	5.56
Ch_B	75.76	94.36	35.64	22.66
Ch_C	49.67	62.69	47.03	35.25
Ch_D	17.25	23.47	65.45	57.11

Table 5. Performance statistics for the proposed multiuser schemes

Mu-Opportunistic and Mu-Threshold are two efficient multiuser schemes but there is still a margin for improvement in order to achieve the upper bound set by the Mu-Ideal. The second metric presented in Table 5 refers to the the available improvement margin. The three schemes share the principle of opportunistic scheduling, but implement it in different ways. Mu-Ideal assumes perfect CSI knowledge without any additional overhead cost, which is an assumption that does not hold for realistic schemes. Mu-Opportunistic introduces considerable overhead since $N = 10$ CTS packets are sent in each transmission sequence. Finally, Mu-Threshold manages to reduce overhead by employing m control slots, with m usually much smaller than the number of total users N (in the presented example, the best performance throughput has been obtained for no more than $m = 3$ slots). As a result, Mu-Threshold is closer to the Mu-Ideal.

Another interesting observation is that the two practical schemes are closer to the ideal under worse channel conditions. In the case of Ch_A, for instance, the improvement margin is 26.8 % for Mu-Opportunistic and only 5.6 % for Mu-Threshold (less that 1 Mbps below the upper throughput bound). The gap between the achieved throughput and the ideal performance opens as the channel conditions improve and in the case of Ch_D both schemes have an improvement margin of more than 50 %. This occurs because the overhead information, consisting of control packets transmitted at the lowest rate, has a greater impact on performance when high data rates are employed.

Table 6 gives an estimation of the improvement achieved by exploiting the multiuser diversity. This gain is reflected in the increase of the average data transmission rate compared to the average user rate for each channel model. The average data transmission rate is calculated as the average of the rates employed for the transmission of all data frames. The average user rate is obtained by calculating the average value of the maximum rate at which a user can transmit, if the best beam (i.e., with the higher SNIR) for the particular user is selected. This value depends on the channel model and is indicated in the second column of the table.

Channel	Avg. User Rate (Mbps)	Avg. Tx Rate (Mbps)			
		Mu-Basic	Mu-Opport.	Mu-Thres.	Mu-Ideal
Ch_A	12	9.73	18.77	27.60	18.77
Ch_B	18	14.37	23.76	30.48	23.76
Ch_C	24	19.01	34.46	44.40	34.46
Ch_D	36	32.41	46.73	51.64	46.73

Table 6. Multiuser diversity gain

In the case of Mu-Basic, the average transmission rate is lower than the average user rate. This is a direct consequence of random scheduling and beam allocation: users may be selected for transmission when their channel quality is low, or they may receive increased interference from other simultaneous transmissions due to the suboptimal beam allocation. Mu-Opportunistic, on the other hand, exploits multiuser diversity by assigning the best user on every beam. As a result, most transmissions take place at rates above the average. In the case of Ch_D, for instance, the transmission rate is 46.7 Mbps whereas the average user rate is limited to 36 Mbps. It should be noted that Mu-Opportunistic yields the same average transmission rate as the Mu-Ideal scheme, since both schemes implement the same scheduling policy. Despite providing the same transmission rate, the throughput performance of Mu-Opportunistic is lower than the ideal, due to the additional control overhead required for the CSI acquisition.

Finally, the maximum transmission rate values are achieved by Mu-Threshold. At first glance, is seems puzzling to obtain rates above those of the Mu-Ideal scheme. Nevertheless, this can be explained with the help of the data presented in Table 4. By imposing a rate threshold, Mu-Threshold scheme controls the minimum rate that can be employed for transmission.

http://dx.doi.org/10.5772/57129

For instance, in the case of Ch_D, the optimum performance is achieved for a threshold of 48 Mbps, meaning that all transmissions have taken place at the rates of 48 and 54 Mbps, thus increasing the average transmission rate. On the other hand, since the average user rate for this channel is 36 Mbps, there is a high possibility that users may not satisfy the threshold condition, resulting to empty frames with no data transmissions. For the best configuration of Mu-Threshold for Ch_D, the percentage of empty frames is approximately 24% of the total frame sequences, as indicated in the last column of Table 4.

So far, a relatively small number of users, $N = 10$ has been considered. The following set of plots in Figure 15 shows the maximum throughput obtained by Mu-Opportunistic and Mu-Threshold as a function of the number of system users N for channel models Ch_A and Ch_D. The best configuration for the Mu-Threshold has been considered and the employed values for the slot number and the rate threshold are also indicated in the figure.

In the case of Mu-Opportunistic, throughput decreases as the number of users grows. This is an unavoidable consequence of the control overhead required for the acquisition of CSI by all the system users. The degradation is more pronounced as the channel improves (e.g., Ch_D) and higher rates are employed for the data transmission (but not the overhead that is sent at the lowest supported rate). The lesson learned from this observation is that when multiple users are present, the Mu-Opportunistic mechanism is not very efficient. As a more viable alternative, the AP could divide the users in smaller groups and poll a user subset in each transmission sequence. This would reduce the multiuser gain but would also limit the introduced overhead.

On the other hand, Mu-Threshold handles multiple users in a more efficient way and throughput is actually improved as the user number increases. Several factors affect this behavior. First, when more users are present, the gain extracted from multiuser diversity also increases, since there is a higher probability of assigning a high-user rate on each beam. Second, unlike the Mu-Opportunistic scheme, the control overhead does not increase linearly with N but depends on the number of CTS slots m.

The selection of the slot number and the rate threshold provides a flexible mechanism to control the number of participating users in each transmission sequence. The best configuration depends on the channel distribution but generally the following principles hold:

- More slots are required as the number of user grows, to reduce the collision probability in the contention window. By observing the best configuration for each case, marked in Figure 15, it can be said that m generally follows an increasing trend.
- The collision probability can also be reduced by increasing the rate threshold, which results to a smaller number of participating users (but with higher available rates). Again, as the number of users grows, the threshold is progressively raised.

6. Conclusions

This chapter has presented a novel approach for the integration of multiuser capabilities in IEEE 802.11n/ac based WLANs. On one hand, a low-complexity beamforming technique named MOB has been employed at the PHY layer. The main strength of MOB lies in the

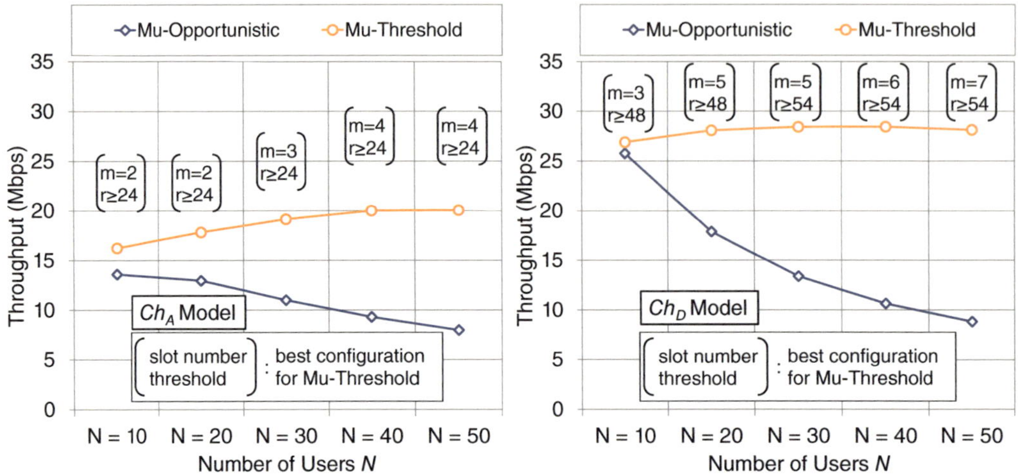

(a) Channel Ch_A (b) Channel Ch_D

Figure 15. Throughput performance comparison versus the number of users

fact that it only requires partial CSI information at the transmitter side, in the form of SNIR measurements acquired by the downlink users. Since the IEEE 802.11n/ac specifications support beamforming, MOB can be easily implemented with minor modifications in the beamforming steering matrices.

On the other hand, in order to exploit the potential of the MOB technique in a realistic scenario, it is necessary to design appropriate MAC layer mechanisms to handle multiuser transmissions. In this chapter, three MAC layer schemes have been proposed. The first scheme, Mu-Basic, implemented a simple random scheduling multiuser scheme, meant to serve as a performance reference. Then, two opportunistic schemes have been proposed, Mu-Opportunistic and Mu-Threshold, that enhance performance by extracting the multiuser diversity gain.

The performance evaluation of the proposed multiuser schemes has led to many interesting observations. The lessons learned can be employed to improve the proposed algorithms but also as more general guidelines in the design of multiuser MAC schemes. The more remarkable conclusions are summarized as follows:

- When the best set of users is opportunistically selected depending on their measured channel quality, the gain achieved by multiuser transmissions can be significant.
- Multiuser diversity gain increases with the number of system users, since there is a higher probability of finding a high-rate set of users among a larger user pool. On the other hand, more users come with a cost of additional control overhead for the CSI acquisition. Mu-Threshold handles efficiently multiple users by setting appropriately the slot number

and the rate threshold parameters. In the case of Mu-Opportunistic scheme, the control overhead increases linearly with the user number and performance eventually drops.

- Under harsh channels, the performance of the proposed multiuser schemes approaches the upper performance bound set by the ideal case of having perfect CSI knowledge with no additional overhead. On the other hand, under more favorable channel conditions, there is still a margin for potential improvement by exploiting multiuser transmissions.

Acknowledgements

This work has been funded by the Research Projects GREENET (PITN-GA-2010-264759), CO2GREEN (TEC2010-20823), Green-T (CP8-006) and GEOCOM (TEC2011-27723-C02-01).

Author details

Elli Kartsakli[1,*], Nizar Zorba[1], Luis Alonso[2] and Christos Verikoukis[3]

* Address all correspondence to: ellik@tsc.upc.edu

1 Signal Theory and Communications Department of Technical University of Catalunya (UPC), Barcelona, Spain
2 Electrical Engineering Department of The University of Jordan, Jordan
3 Telecommunications Technological Center of Catalunya (CTTC), Barcelona, Spain

References

[1] IEEE Standard for Information technology – Telecommunications and information exchange between systems – Local and metropolitan area networks – Specific requirements – Part 11: Wireless LAN Medium Access Control (MAC) and Physical Layer (PHY) specifications. *IEEE Std 802.11-2007 (Rev. of IEEE Std 802.11-1999)*, December 2007.

[2] IEEE Standard for Information Technology – Telecommunications and information exchange between systems – Local and metropolitan area networks – Specific requirements Part 11: Wireless LAN Medium Access Control (MAC) and Physical Layer (PHY) specifications Amendment 5: Enhancements for Higher Throughput. *IEEE Std 802.11n-2009*, pages c1–502, October 2009.

[3] G. Caire and S. Shamai. On the achievable throughput of a multiantenna Gaussian broadcast channel. *IEEE Transactions on Information Theory*, 49(7):1691–1706, July 2003.

[4] M.Z. Siam and M. Krunz. An overview of MIMO-oriented channel access in wireless networks [medium access control protocols for wireless LANs]. *IEEE Wireless Communications Magazine*, 15(1):63–69, February 2008.

[5] Per H. Lehne and Magne Pettersen. An Overview of Smart Antenna Technology for Mobile Communications Systems. *IEEE Communication Surveys &Tutorials*, 2(4):2–13, 4th Quarter 1999.

[6] M. Zorzi, J. Zeidler, A. Anderson, B. Rao, J. Proakis, A.L. Swindlehurst, M. Jensen, and S. Krishnamurthy. Cross-layer issues in MAC protocol design for MIMO ad hoc networks. *IEEE Wireless Communications Magazine*, 13(4):62 – 76, August 2006.

[7] D. Gesbert, M. Shafi, Da shan Shiu, P.J. Smith, and A. Naguib. From theory to practice: an overview of MIMO space-time coded wireless systems. *IEEE Journal on Selected Areas in Communications*, 21(3):281–302, April 2003.

[8] W. C. Jakes. *Microwave Mobile Communications*. Wiley and Sons, New York, NY, 1974.

[9] N. Seshadri and J.H. Winters. Two signaling schemes for improving the error performance of frequency-division-duplex (FDD) transmission systems using transmitter antenna diversity. In *Proc. of IEEE 43rd Vehicular Technology Conference (VTC 1993)*, pages 508 – 511, May 1993.

[10] A. Wittneben. Basestation modulation diversity for digital simulcast. In *Proc. of IEEE 41th Vehicular Technology Conference (VTC 1991)*, pages 848–853, May 1991.

[11] A. Wittneben. A new bandwidth efficient transmit antenna modulation diversity scheme for linear digital modulation. In *Proc. of IEEE International Conference on Communications (ICC 1993)*, pages 1630 – 1634 vol.3, May 1993.

[12] S.M. Alamouti. A simple transmit diversity technique for wireless communications. *IEEE Journal on Selected Areas in Communications*, 16(8):1451–1458, October 1998.

[13] V. Tarokh, N. Seshadri, and A.R. Calderbank. Space-time codes for high data rate wireless communication: Performance criterion and code construction. *IEEE Transactions on Information Theory*, 44(2):744–765, March 1998.

[14] V. Tarokh, H. Jafarkhani, and A.R. Calderbank. Space-time block codes from orthogonal designs. *IEEE Transactions on Information Theory*, 45(5):1456 – 1467, July 1999.

[15] P.W. Wolniansky, G.J. Foschini, G.D. Golden, and R.A. Valenzuela. V-BLAST: an architecture for realizing very high data rates over the rich-scattering wireless channel. In *URSI International Symposium on Signals, Systems, and Electronics (ISSSE 1998)*, pages 295–300, September 1998.

[16] A.J. Paulraj and C.B. Papadias. Space-time processing for wireless communications. *IEEE Signal Processing Magazine*, 14(6):49–83, November 1997.

[17] Lizhong Zheng and D.N.C. Tse. Diversity and multiplexing: a fundamental tradeoff in multiple-antenna channels. *IEEE Transactions on Information Theory*, 49(5):1073 – 1096, May 2003.

[18] D.N.C. Tse, P. Viswanath, and Lizhong Zheng. Diversity-multiplexing tradeoff in multiple-access channels. *IEEE Transactions on Information Theory*, 50(9):1859–1874, September 2004.

[19] C. Anton-Haro, P. Svedman, M. Bengtsson, A. Alexiou, and A. Gameiro. Cross-layer scheduling for multi-user MIMO systems. *IEEE Communications Magazine*, 44(9):39–45, September 2006.

[20] A.S. Macedo and E.S. Sousa. Antenna-sector time-division multiple access for broadband indoor wireless systems. *IEEE Journal on Selected Areas in Communications*, 16(6):937–952, August 1998.

[21] Hujun Yin and Hui Liu. Performance of space-division multiple-access (SDMA) with scheduling. *IEEE Transactions on Wireless Communications*, 1(4):611–618, October 2002.

[22] D. Gesbert and M.-S. Alouini. How much feedback is multi-user diversity really worth? In *Proc. of IEEE International Conference on Communications (ICC 2004)*, pages 234–238, June 2004.

[23] S. Sanayei and A. Nosratinia. Opportunistic Downlink Transmission With Limited Feedback. *IEEE Transactions on Information Theory*, 53(11):4363–4372, November 2007.

[24] Xiangping Qin and R. Berry. Opportunistic splitting algorithms for wireless networks. In *Proc. of the 23rd Annual Joint Conference of the IEEE Computer and Communications Societies (INFOCOM 2004)*, pages 1662–1672, vol.3, 2004.

[25] V. Shah, N.B. Mehta, and R. Yim. Analysis, Insights and Generalization of a Fast Decentralized Relay Selection Mechanism. In *Proc. of IEEE International Conference on Communications (ICC 2009)*, pages 1–6, June 2009.

[26] J. Mirkovic, Jing Zhao, and D. Denteneer. A MAC Protocol with Multi-User MIMO Support for Ad-Hoc WLANs. In *Proc. of IEEE 18th International Symposium on Personal, Indoor and Mobile Radio Communications (PIMRC 2007)*, pages 1–5, September 2007.

[27] Young-June Choi, Neung-Hyung Lee, and Saewoong Bahk. Exploiting Multiuser MIMO in the IEEE 802.11 Wireless LAN Systems. *Wireless Personal Communications*, 54(3):385–396, August 2008.

[28] V. Shah, N.B. Mehta, and R. Yim. MAC enhancements for downlink multi-user MIMO transmission in next generation WLAN. In *Proc. of IEEE Consumer Communications and Networking Conference (CCNC 2012)*, pages 832–837, January 2012.

[29] IEEE 802 11-03/161r2. TGn Indoor MIMO WLAN Channel Models, 2004.

[30] M. Sharif and B. Hassibi. On the capacity of MIMO broadcast channels with partial side information. *IEEE Transactions on Information Theory*, 51(2):506–522, February 2005.

[31] Nizar Zorba and A.I. Pérez-Neira. CAC for Multibeam Opportunistic Schemes in Heterogeneous WiMax Systems Under QoS Constraints. In *Proc. of IEEE Global Telecommunications Conference (GLOBECOM 2007)*, pages 4296–4300, November 2007.

[32] P. Viswanath, D.N.C. Tse, and R. Laroia. Opportunistic beamforming using dumb antennas. *IEEE Transactions on Information Theory*, 48(6):1277–1294, June 2002.

[33] Elli Kartsakli, Nizar Zorba, Luis Alonso, and Christos Verikoukis. A threshold-selective multiuser downlink mac scheme for 802.11n wireless networks. *IEEE Transactions on Wireless Communications*, 10(3):857 –867, March 2011.

[34] D. Pubill and A.I. Pérez-Neira. Handoff Optimization with Fuzzy Logic in 802.11 Networks. In *Proc. of Conference on Information Processing and Management of Uncertainty in Knowledge-Based Systems (IPMU 2006)*, July 2006.

Section 6

Physical Layer Security for Multi-User MIMO Communications

Chapter 6

Physical Layer Security for Multiuser MIMO Communications

Giovanni Geraci and Jinhong Yuan

Additional information is available at the end of the chapter

http://dx.doi.org/10.5772/57130

1. Introduction

Wireless multi-user MIMO communications are used more and more often to exchange sensitive data. Because of the broadcast nature of the physical medium, unauthorized receivers located within the transmission range can observe the signals sent by the transmitter to a legitimate receiver and eavesdrop them. Therefore, security has become an extremely important issue to deal with. Multiuser MIMO communications are particularly sensitive to the problem of security, because each confidential message must be kept secret not only from external nodes, but also from all the users other than the intended one.

Traditionally, wireless security is ensured by network-layer cryptography techniques. However, these techniques may not be suitable in the case of large dynamic wireless networks, since they raise issues like key distribution and management (for symmetric cryptosystems), and high computational complexity (for asymmetric cryptosystems). Moreover, these schemes are potentially vulnerable, since they rely on the limited resources of the eavesdropper and on the unproven assumption that certain encryption algorithms are hard to invert. Methods exploiting the randomness inherent in noisy channels, known as physical layer security, have been proposed to enhance the protection of transmitted data and achieve perfect secrecy [1, 2]. Physical-layer security allows secret communications over a wireless channel without requiring an encryption key, and it works by limiting the amount of information that can be extracted at the physical level by an unintended receiver. This is performed by designing appropriate coding and precoding schemes, and by exploiting the channel state information available at the network nodes [3].

Physical layer security for communications was proposed in the 1970's by Wyner [4], who studied a three-terminal network consisting of a transmitter, an intended user and an eavesdropper, known as the wiretap channel. For this network, the secrecy capacity was defined as the maximum rate at which a message can be transmitted reliably to the intended user while the rate of information leakage to the eavesdropper vanishes asymptotically with the code length. For the case when the eavesdropper's channel is a degraded version of the intended user's channel, Wyner showed that it is possible to have secret communication without using an encryption key. This can be achieved by a randomized coding scheme where the information is hidden in the additional noise seen by the eavesdropper. Each message is mapped to many codewords, thus inducing maximal equivocation at the eavesdropper. Csizar and Korner generalized Wyner's work by considering a nondegraded version of the wiretap channel [5].

Physical layer security was then applied to Gaussian channels [6], and it was observed that a secret transmission can be achieved only if the channel at the eavesdropper is noisier than the channel at the intended user. The presence of slow fading was shown to significantly change the situation, since it allows the transmitter to employ a variable-rate transmission, thus achieving secrecy even when the eavesdropper's channel is better than the intended receiver's channel on average [7]. Also the use of multiple antennas can enhance the secrecy capability, because it enables the transmitter to beamform in a direction as orthogonal to the eavesdropper and as close to the intended user as possible [8–10]. Even when the channel at the eavesdropper is unknown by the transmitter, artificial noise can be transmitted to degrade the eavesdropper's channel and thus reduce its signal-to-noise ratio, while being harmless to the intended receiver [11–13].

More recently, physical layer security has also been extended to multiuser MIMO channels. In this chapter, we will survey the research in the field of physical layer security for multiuser MIMO communications, especially focusing on the case when multiple malicious users are present in the network, and they can eavesdrop on each other. For these complex scenarios, we will present some suboptimal low-complexity transmission schemes, discuss their performance and quantify the sum-rate penalties imposed by the secrecy requirements and by the presence of multiple users. We will discuss the challenges that arise in networks with a large number of malicious receivers, we will identify potential ways to deal with these challenges, and present an outlook on future directions for research.

2. Physical layer security in multiuser MIMO systems

One way to extend the concept of physical layer security to multiuser systems is by considering the multiuser wiretap channel, where a transmitter wants to have confidential communication with an arbitrary number of trusted users in the presence of an external eavesdropper. For this system set-up, the secrecy capacity region in the presence of an arbitrary number of legitimate receivers was characterized in [14], by using the relationship between the minimum-mean-square-error and the mutual information. The capacity achieving coding scheme was shown to be a variant of dirty-paper coding with Gaussian signals.

Since the transmitter cannot always predict the behavior of the users, the multiuser MIMO channel with malicious users is now regarded with large interest. This is also denoted as the broadcast channel with confidential messages (BCC). Consider a broadcast channel with two independent confidential messages sent to two receivers, where each receiver acts as an eavesdropper for the other one. In other words, the first message is intended for

http://dx.doi.org/10.5772/57130

the first receiver but needs to be kept secret from the second receiver, and viceversa. This scenario was studied in [15] for the multiple-input single-output (MISO) Gaussian case and in [16] for general MIMO Gaussian case. In this case it was shown that both confidential messages can be simultaneously transmitted at their respective maximum secrecy rates, and the achievability was obtained using the dirty-paper coding.

Let us cosider now a larger multiuser network with more than two malicious users. For this network, it is required that the base station (BS) securely transmits each confidential message, ensuring that none of the other unintended users receive any information. Since in general the behavior of the users cannot be determined by the transmitter, a conservative worst-case scenario can be assumed for each user, where all the remaining users can cooperate to jointly eavesdrop. In this case, for each user, the alliance of the cooperating eavesdropper is equivalent to a single multi-antenna eavesdropper.

The MISO BCC with a generic number of malicious receivers was studied in [17, 18], and it consists of a BS with M antennas that simultaneously transmit independent confidential messages to K spatially dispersed single-antenna users, which can cooperate and eavesdrop on each other. Although determining the secrecy capacity region for the generic MISO BCC is still an open problem, suboptimal transmission schemes have been proposed to achieve high secrecy sum-rates by controlling the amount of crosstalk between the users [19]. These schemes are based on linear precoding, and unlike dirty-paper coding, their low complexity makes them suitable for practical implementation. In the following sections, we present some new results on the secrecy sum-rates achieved by multiuser MIMO linear precoding.

3. Physical layer security with multi-user MIMO linear precoding

Although suboptimal, linear precoding schemes are of particular interest because of their low-complexity implementations and because they can control the amount of crosstalk between the users to maintain a high sum-rate in the broadcast channel [20–27]. In the MISO BCC, linear precoding can be employed to control the amount of interference and information leakage to the unintended receivers introduced by the transmission of each confidential message [17–19].

Let the transmitted signal be denoted by $\mathbf{x}$, then the received signal is given by

$$\mathbf{y} = \mathbf{H}\mathbf{x} + \mathbf{n} \tag{1}$$

where $\mathbf{H} = [\mathbf{h}_1, \ldots, \mathbf{h}_K]$ is the $K \times M$ channel matrix, $\mathbf{h}_k$ is the k-th column of $\mathbf{H}$ and it represents the channel between the BS and the k-th user, and $\mathbf{n}$ is complex Gaussian noise. In linear precoding, the transmitted vector $\mathbf{x}$ is derived from the vector containing the confidential messages $\mathbf{u} = [u_1, \ldots, u_K]^T$ through a deterministic linear transformation (precoding) [22–25]. Let $\mathbf{W} = [\mathbf{w}_1, \ldots, \mathbf{w}_K]$ be the $M \times K$ precoding matrix, where $\mathbf{w}_k$ is the k-th column of $\mathbf{W}$. Then the transmitted signal is

$$\mathbf{x} = \mathbf{W}\mathbf{u} = \sum_{k=1}^{K} \mathbf{w}_k u_k. \tag{2}$$

3.1. Achievable secrecy sum-rates with linear precoding

The secrecy sum-rates achievable by linear precoding were obtained in [18] by considering the worst-case scenario, where for each intended receiver k the remaining $K-1$ users can form an alliance $\tilde{k}$, and cooperate to jointly eavesdrop on the message u_k. By noting that each user k, along with its own eavesdropper $\tilde{k}$ and the transmitter, forms an equivalent multi-input, single-output, multi-eavesdropper (MISOME) wiretap channel [10], an achievable secrecy sum-rate R_s is given by

$$R_s = \sum_{k=1}^{K} \max\left\{\log_2\left(1+\mathrm{SINR}_k\right) - \log_2\left(1+\mathrm{SINR}_{\tilde{k}}\right), 0\right\}, \tag{3}$$

where SINR_k and $\mathrm{SINR}_{\tilde{k}}$ are the signal-to-interference-plus-noise ratios for the message u_k at the intended receiver k and the eavesdropper $\tilde{k}$, respectively, given by

$$\mathrm{SINR}_k = \frac{\rho\left|\mathbf{h}_k^H\mathbf{w}_k\right|^2}{1+\rho\sum_{j\neq k}\left|\mathbf{h}_k^H\mathbf{w}_j\right|^2} \tag{4}$$

and

$$\mathrm{SINR}_{\tilde{k}} = \rho\left\|\mathbf{H}_k\mathbf{w}_k\right\|^2, \tag{5}$$

and where ρ is the transmit SNR, and $\mathbf{H}_k$ is the matrix obtained from $\mathbf{H}$ by removing the k-th row.

Particular attention was given to the Regularized Channel Inversion (RCI) precoder, because it achieves better performance than the plain Channel Inversion precoder, especially at low SNR [24, 25]. A linear precoder based on RCI was proposed for the MISO BCC in [19]. The RCI precoding matrix is given by

$$\mathbf{W} = \frac{1}{\sqrt{\gamma}}\mathbf{H}^H\left(\mathbf{H}\mathbf{H}^H + M\xi\mathbf{I}_K\right)^{-1}, \tag{6}$$

where γ is a long-term power normalization constant, given by

$$\gamma = \mathrm{tr}\left\{\mathbf{H}^H\mathbf{H}(\mathbf{H}^H\mathbf{H} + M\xi\mathbf{I}_M)^{-2}\right\}. \tag{7}$$

For each message, the function of the regularization parameter ξ is to achieve a tradeoff between maximizing the signal power at the intended user and minimizing the interference and information leakage at the other unintended users. In [19], the regularization parameter is optimized to maximize the secrecy sum-rate.

3.2. Large-system results

The secrecy sum-rate achievable by the RCI precoder in the MISO BCC was obtained in [19] by large-system analysis, where both the number of receivers K and the number of transmit antennas M approach infinity, with their ratio $\beta = K/M$ being held constant. Unless otherwise stated, the results presented in the following refer to the large-system regime. We note that these results are accurate even when applied to small systems with a finite number of users.

An expression for the secrecy sum-rate R_s° in the large-system regime is given by [19]

$$R_s^\circ = \max\left\{ K\log_2 \frac{1+g(\beta,\xi)\,\frac{\rho+\frac{\rho\xi}{\beta}[1+g(\beta,\xi)]^2}{\rho+[1+g(\beta,\xi)]^2}}{1+\frac{\rho}{(1+g(\beta,\xi))^2}}, 0\right\} \tag{8}$$

with

$$g(\beta,\xi) = \frac{1}{2}\left[\operatorname{sgn}(\xi)\cdot\sqrt{\frac{(1-\beta)^2}{\xi^2}+\frac{2(1+\beta)}{\xi}+1}+\frac{1-\beta}{\xi}-1\right]. \tag{9}$$

In [19], a closed form expression was also derived for the optimal regularization parameter $\xi^{\star\circ}$, given by

$$\xi^{\star\circ} = \frac{-2\rho^2(1-\beta)^2+6\rho\beta+2\beta^2-2[\beta(\rho+1)-\rho]\cdot\sqrt{\beta^2[\rho^2+\rho+1]-\beta[2\rho(\rho-1)]+\rho^2}}{6\rho^2(\beta+2)+6\rho\beta}. \tag{10}$$

For the specific case $\beta = 1$, i.e. $M = K$, the value of $\xi^{\star\circ}$ reduces to [18]

$$\xi^{\star\circ} = \frac{1}{3\rho+1+\sqrt{3\rho+1}}, \quad \text{for } \beta = 1. \tag{11}$$

We note that the value of the regularization parameter $\xi^{\star\circ}$ that maximizes the secrecy sum-rate differs from the value $\xi^{\star\circ}_{\text{ns}} = \beta/\rho$ that maximizes the sum-rate without secrecy requirements [28].

By substituting the optimal value of the regularization parameter (10) in (8), it is possible to obtain the optimal secrecy sum-rate $R_s^{\star\circ}$ achievable by RCI precoding in the large-system regime. The secrecy sum-rate $R_s^{\star\circ}$ is a function of K, β and ρ. When $\beta = 1$, the optimal secrecy sum-rate $R_s^{\star\circ}$ has a simple expression, given by

$$R_s^{\star\circ} = K \log_2 \frac{9\rho + 2 + (6\rho + 2)\sqrt{3\rho + 1}}{4(4\rho + 1)}, \quad \text{for } \beta = 1. \tag{12}$$

Although the optimal value of the regularization parameter $\xi^{\star\circ}$ in (10) was derived in [19] in the large-system regime, using $\xi^{\star\circ}$ in a finite-size system does not cause a significant loss in the secrecy sum-rate compared to using a regularization parameter $\xi_{\mathrm{fs}}(\mathbf{H})$ optimized for every channel realization.

Fig. 1 shows the complementary cumulative distribution function (CCDF) of the normalized secrecy sum-rate difference between using $\xi^{\star\circ}$ and $\xi_{\mathrm{fs}}(\mathbf{H})$ as the regularization parameter of the RCI precoder for $K = 4, 8, 16, 32$ users, for $\beta = 1$ and at an SNR of 10dB. The difference is normalized by dividing by the secrecy sum-rate of the precoder that uses $\xi_{\mathrm{fs}}(\mathbf{H})$. We observe that the average normalized secrecy sum-rate difference is less than 2.4 percent for all values of K. As a result, the large-system regularization parameter $\xi^{\star\circ}$ may be used instead of the finite-system regularization parameter with only a small loss of performance. Moreover, the value of $\xi^{\star\circ}$ does not need to be calculated for each channel realization.

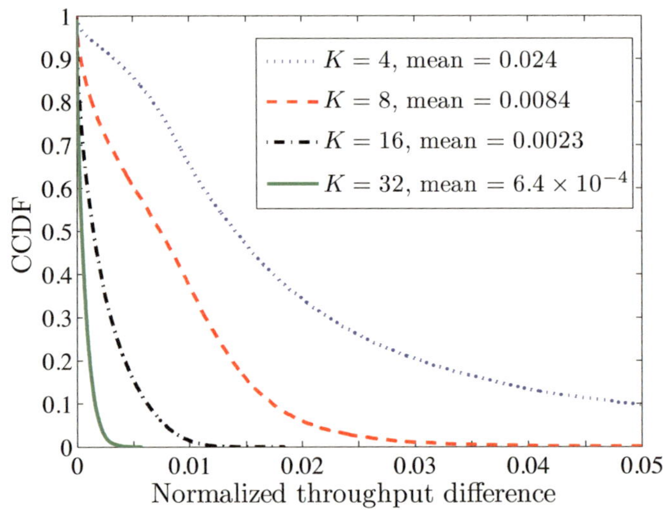

Figure 1. Complementary cumulative distribution function (CCDF) of the normalized secrecy sum-rate difference between using $\xi_{\mathrm{fs}}(\mathbf{H})$ and $\xi^{\star\circ}$, with $\beta = 1$ and $\rho = 10$dB.

Fig. 2 compares the secrecy sum-rate $R_s^{\star\circ}$ of the RCI precoder from the large-system analysis to the simulated ergodic secrecy sum-rate R_s with a finite number of users, for different values of β. We observe that as M increases, the simulated rates approach the curves from large-system analysis. For $\beta \leq 1$, $R_s^{\star\circ}$ is always positive and monotonically increasing with the SNR ρ. However when $\beta > 1$, the secrecy sum-rate does not monotonically increase with ρ. There is an optimal value of the SNR beyond which the achievable secrecy sum-rate $R_s^{\star\circ}$ starts decreasing, until it becomes zero for large SNR. When $\beta \geq 2$ no positive secrecy sum-rate is achievable at all [19].

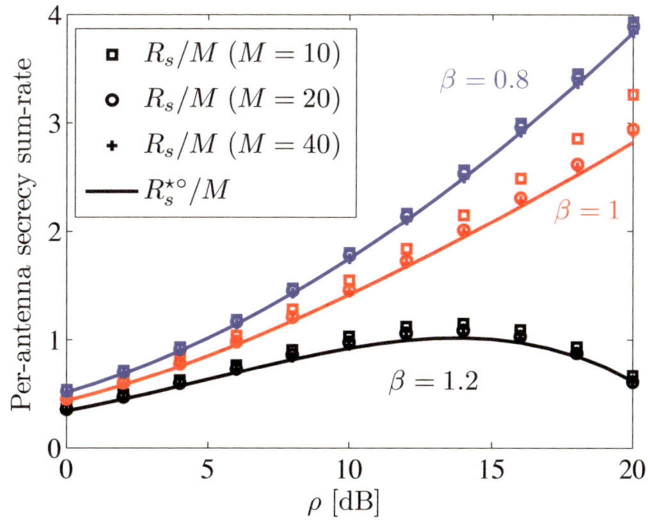

Figure 2. Comparison between the secrecy sum-rate with RCI precoding in the large-system regime (8) and the simulated ergodic secrecy sum-rate for finite M. Three sets of curves are shown, each one corresponds to a different value of β.

The expression of the secrecy sum-rate $R_s^{\star\circ}$ becomes simpler in the limit of large SNR. In fact, it can be approximated by

$$R_s^{\star\circ} \approx \begin{cases} K \log_2 \frac{1-\beta}{\beta} + K \log_2 \rho & \text{for } \beta < 1 \\ \frac{K}{2} \log_2 \frac{27}{64} + \frac{K}{2} \log_2 \rho & \text{for } \beta = 1 \\ \max\left\{3K \log_2 \frac{\beta}{\beta - 1} - K \log_2 \rho, 0\right\} & \text{for } \beta > 1 \end{cases}, \quad \text{as } \rho \to \infty. \tag{13}$$

We note from (13) that for high SNR, the behavior of the secrecy sum-rate significanly depends on the ratio β between the number of users K and the number of transmit antennas M. When $K < M$, the secrecy sum-rate scales linearly with the factor K. If $K = M$, the scaling factor reduces to $K/2$. When the number of users K exceeds the number of antennas M, then the secrecy sum-rate decreases with the SNR ρ, and there is a value of ρ beyond which the achievable secrecy sum-rate becomes zero.

3.3. Effect of the network load

Fig. 3 depicts the asymptotic secrecy sum-rate per transmit antenna as a function of β, for several values of the SNR. We denote by β_{opt} the value of the ratio β that maximizes the secrecy sum-rate per transmit antenna $R_s^{\star\circ}/M$. It is possible to see from Fig. 3 that β_{opt} is an increasing function of the SNR. The value of β_{opt} falls between 0 and 1, and it tends to 1 in the limit of large SNR.

We denote by $\beta_{\max}$ the maximum value of β allowed for non-zero secrecy sum-rates. The value of $\beta_{\max}$ represents the maximum number of users per transmit antenna that can be served with non-zero secrecy sum-rate. Fig. 3 shows that $\beta_{\max}$ is a decreasing function of the SNR. The value of $\beta_{\max}$ can be found by solving the following cubic equation [19]

$$(\rho+1)\,\beta_{\max}^3 - (3\rho+2)\,\beta_{\max}^2 + 3\rho\beta_{\max} - \rho = 0. \tag{14}$$

The value of $\beta_{\max}$ falls between 1 and 2. This means that if $K \geq 2M$, i.e. if $\beta \geq 2$, then the secrecy sum-rate is zero for all SNRs. In the limit of large SNR, equation (14) reduces to

$$(\beta_{\max} - 1)^3 = 0, \tag{15}$$

yielding to $\beta_{\max} = 1$. These results can be explained as follows. In the worst-case scenario, the alliance of cooperating eavesdroppers can cancel the interference, and its received SINR is the ratio between the signal leakage and the thermal noise. In the limit of large SNR, the thermal noise vanishes, and the only means for the transmitter to limit the eavesdropper's SINR is by reducing the signal leakage to zero by inverting the channel matrix. This can only be accomplished when the number of transmit antennas is larger than or equal to the number of users, hence only if $\beta \leq 1$. When $\beta > 1$ this is not possible, and no positive secrecy sum-rate can be achieved. When $\beta \geq 2$, the eavesdroppers are able to drive the secrecy sum-rate to zero irrespective of ρ. This is consistent with the results presented in [10] for a single-user system.

4. The cost of physical layer security in multi-user MIMO

Guaranteeing secrecy and serving multiple (and potentially malicious) users at the same time both come at a cost in terms of the per-user transmission rate. In this section, we discuss the cost of achieving physical layer security in multiuser MIMO communications.

4.1. Secrecy loss

The cost due to the secrecy requirements, which we denote by *secrecy loss*, can be obtained by comparing the secrecy sum-rate $R_s^{\star\circ}$ achieved by the RCI precoder to the sum-rate $R^{\star\circ}$ achieved by an optimized RCI precoder without secrecy requirements. The gap between $R_s^{\star\circ}$ and $R^{\star\circ}$ represents how much guaranteeing secrecy costs in terms of the achievable sum-rate.

The optimal sum-rate $R^{\star\circ}$ without secrecy requirements is obtained by using the precoder in (6), and it is given by [29]

$$R^{\star\circ} = K \log_2 \left[1 + g\left(\beta, \xi_{\rm ns}^{\star\circ}\right)\right], \tag{16}$$

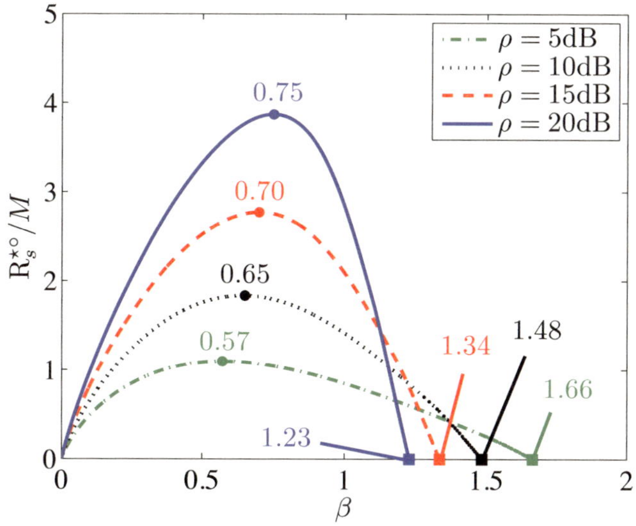

Figure 3. Asymptotic secrecy sum-rate per transmit antenna with RCI as a function of β. Circles denote β_{opt}, squares denote $\beta_{\max}$.

with $\xi_{\text{ns}}^{\star\circ} = \beta/\rho$. It is easy to show that $R^{\star\circ} \geq 0$ for all values of β and ρ, with equality only for $\rho = 0$, and that $R^{\star\circ}$ tends to zero as $\beta \to \infty$. Hence, there is no limit to the number of users per transmit antenna β that the system can accommodate with a non-zero sum-rate. However if we impose the secrecy requirements, the secrecy sum-rate $R_s^{\star\circ}$ is zero for $\beta \geq \beta_{\max}$, with $\beta_{\max}$ given by (14). Therefore, introducing the secrecy requirements will limit to $\beta_{\max}$ the number of users per transmit antenna that can be served with a non-zero sum-rate.

We now compare the secrecy sum-rate $R_s^{\star\circ}$ to the sum-rate $R^{\star\circ}$ in the limit of large SNR. Again by using the regularization parameter $\xi_{\text{ns}}^{\star\circ} = \beta/\rho$ we obtain the following large-SNR approximation for the secrecy sum-rate without secrecy requirements [19]

$$R^{\star\circ} \approx \begin{cases} K\log_2 \frac{1-\beta}{\beta} + K\log_2 \rho & \text{for } \beta < 1 \\ \frac{K}{2}\log_2 \rho & \text{for } \beta = 1 \\ K\log_2 \frac{\beta}{\beta-1} & \text{for } \beta > 1 \end{cases}, \quad \text{as } \rho \to \infty. \tag{17}$$

By comparing (17) to (13), we can draw the following conclusions regarding the large-SNR regime. If the number of transmit antennas M is larger than the number of users K, then $\beta < 1$, $R_s^{\star\circ} = R^{\star\circ}$, and the secrecy requirements do not decrease the sum-rate of the network. Therefore, by using $\xi^{\star\circ}$ from (10) one can achieve secrecy while maintaining the same sum-rate, i.e. there is no secrecy loss. If $M = K$, then $\beta = 1$, the secrecy requirements only reduce the sum-rate by a constant value, and the scaling factor $K/2$ remains unchanged. Alternatively, one can achieve secrecy while maintaining the same sum-rate, by increasing the transmitted power by a factor $64/27 \approx 3.75$dB. If $M < K$, i.e. $\beta > 1$, then the secrecy requirements result in a value of $R_s^{\star\circ}$ that decreases with the SNR, as opposed to a constant sum-rate $R^{\star\circ}$ without secrecy. Therefore if the SNR is too large, then the secrecy sum-rate becomes zero.

4.2. Multiuser Loss

The cost due the interference caused by the presence of multiple users in the system, which we denote by *multiuser loss*, is given by the gap between the per-user secrecy rate $R_s^{\star\circ}/K$ and the secrecy capacity $C_{s,\mathrm{SU}}$ of the single-user MISOME wiretap channel, where one user is served at a time and the remaining users can eavesdrop.

The value of $C_{s,\mathrm{SU}}$ was obtained in [10], and for large SNR it can be approximated by

$$C_{s,\mathrm{SU}} \approx \begin{cases} \log_2 \rho & \text{for } \beta < 1 \\ \frac{1}{2}\log_2 \rho & \text{for } \beta = 1 \\ \max\left\{\log_2 \frac{1}{(\beta-1)}, 0\right\} & \text{for } \beta > 1 \end{cases}, \quad \text{as } \rho \to \infty. \tag{18}$$

We compare $R_s^{\star\circ}/K$ to $C_{s,\mathrm{SU}}$ in the large-SNR regime. We note that in $C_{s,\mathrm{SU}}$ from [10] a single-user system is considered. Therefore, only one message is transmitted to one legitimate user, and the user does not experience any interference. By comparing (18) to $R_s^{\star\circ}/K$, we can conclude that for $\beta \leq 1$, the RCI precoder achieves a per-user secrecy rate which has the same linear scaling factor as the secrecy capacity of a single-user system with no interference. When $1 < \beta < 2$, the presence of interference results in a value of $R_s^{\star\circ}$ that decreases with the SNR, as opposed to a constant value for $C_{s,\mathrm{SU}}$. When $\beta \geq 2$, the secrecy rate is zero irrespective of the presence of interference.

4.3. Power allocation

In some cases, the rate loss generated by the secrecy requirements and by the interference due to the presence of multiple users can be compensated by a power allocation scheme. In [18], an iterative power allocation algorithm was proposed to obtain the maximum secrecy sum-rate for a fixed regularization parameter ξ. The algorithm was also extended to maximize the secrecy sum-rate by jointly optimizing the regularization parameter ξ and the power allocation vector. However, in many cases there is a negligible performance difference between the joint and the separate optimization. As a result, a low-complexity, near-optimal RCI precoder may be implemented by using $\xi^{\star\circ}$ from (10) and optimizing the power vector separately [18].

The RCI precoder with optimal power allocation (RCI-PA) outperforms the RCI precoder with equal power (RCI-EP), and the gain does not vanish at high SNR. The RCI-PA precoder thus reduces the rate loss due to secrecy requirements and interference, and in some cases it achieves a per-user rate which is as high as the rate achieved by the optimal RCI-EP precoder without secrecy requirements, and as high as the secrecy capacity of a single-user system [18].

http://dx.doi.org/10.5772/57130

4.4. Numerical results

Fig. 4 compares the simulated ergodic sum-rates R_s and R of the RCI precoder with and without secrecy requirements, respectively. These were obtained by using the regularization parameters $\xi^{\star\circ}$ and $\xi^{\star\circ}_{\text{ns}}$, respectively. For $\beta < 1$, the difference between R and R_s becomes negligible at large SNR, and secrecy can be achieved without additional costs. For $\beta = 1$, the two curves tend to have same slope at large SNR, but there is a residual gap between them. Therefore, secrecy can be achieved at a lower sum-rate. We note that in order to achieve secrecy without decreasing the sum-rate, the required additional power is less than 4dB at all SNRs. For $\beta > 1$, the sum-rate tends to saturate for large SNR, whereas the secrecy sum-rate starts decreasing. If the SNR is too large, then the secrecy requirements force the sum-rate to zero.

Fig. 4 also shows the simulated secrecy capacity $C_{s,\text{SU}}$ of the MISOME wiretap channel. For $\beta \leq 1$, the RCI precoder achieves a per-user secrecy rate which has the same linear scaling factor as $C_{s,\text{SU}}$. When $1 < \beta < 2$, $C_{s,\text{SU}}$ saturates at high SNR, while the secrecy sum-rate decreases. All these numerical results confirm the ones obtained from the large-system analysis.

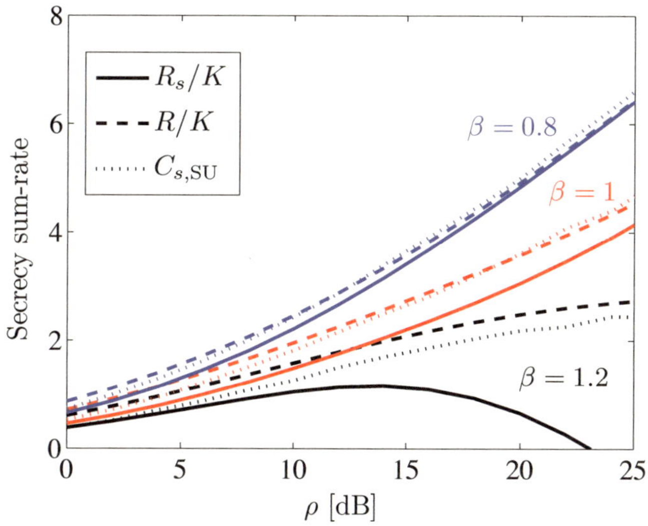

Figure 4. Comparison between the simulated ergodic per-user secrecy rate with RCI (solid) and the two upper bounds: (i) per-user rate without secrecy requirements (dashed) and (ii) MISOME secrecy capacity (dotted), for $K = 12$ users. Three values of β are considered: 0.8, 1, and 1.2, corresponding to $M = 15$, 12 and 10 antennas.

Fig. 5 shows the simulated per-user secrecy rate of the RCI-PA precoder from [18], with optimal power allocation. This is compared to the RCI-EP precoder. Fig. 5 also shows that the power allocation scheme reduces the sum-rate loss due to the secrecy requirements. For $\rho \geq 15$dB, RCI with power allocation achieves a per-user secrecy rate which is even higher than the per-user rate achieved by the optimal RCI-EP without secrecy requirements. Furthermore, Fig. 5 shows the simulated secrecy capacity $C_{s,\text{SU}}$ of a MISOME channel with the same per-message transmit power. Although $C_{s,\text{SU}}$ is obtained in a single-user and interference-free system [10], at high SNR, RCI with power allocation achieves a per-user secrecy rate as large as $C_{s,\text{SU}}$.

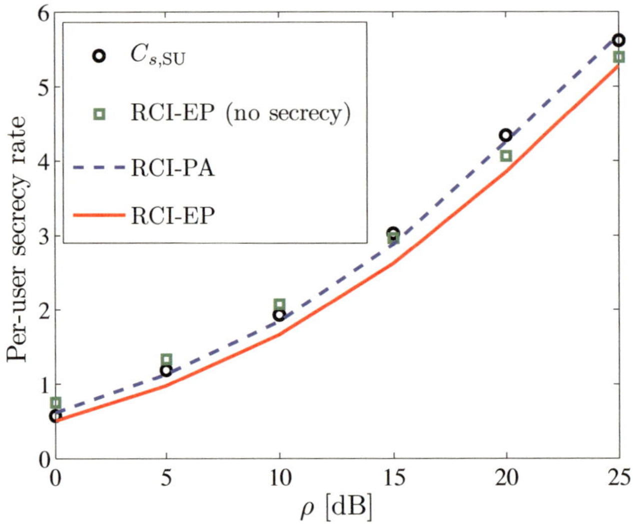

Figure 5. Per-user secrecy rate vs. ρ for $\beta = 1$ and $K = 4$ users: with equal power allocation (solid) and with optimal power allocation (dashed). The rate of the optimal RCI precoder without secrecy requirements (squares) and the secrecy capacity of the MISOME channel (circles) are also plotted.

5. Current research on multiuser MIMO physical layer security

Before concluding this chapter, we briefly discuss current research topics on physical layer security for multiuser MIMO communications, and we mention possible extensions of the results presented.

5.1. Power reduction strategy

Since for $\beta > 1$ the RCI precoder performs poorly in the high-SNR regime, a linear precoder based on RCI and power reduction could significantly increase the high-SNR secrecy sum-rate. In fact, we can observe from Fig. 2 that when $\beta > 1$ there is an optimal value of the SNR beyond which the achievable secrecy sum-rate $R_s^{\star\circ}$ starts decreasing.

A power reduction strategy would prevent the secrecy sum-rate from decreasing at high SNR by reducing the transmit power, and therefore reducing the SNR to the value that maximizes the secrecy sum-rate. For $1 < \beta < 2$ and large SNR, the RCI precoder with power reduction would thus achieve a constant nonnegative secrecy sum-rate. However, this strategy would not be effective for $\beta \geq 2$, since in this case the secrecy sum-rate is zero irrespective of the SNR.

5.2. Secrecy sum-rates in the presence of channel estimation error

In Sections 3 and 4, we discussed the secrecy rate performance of multi-user MIMO linear precoding for the case when perfect channel state information (CSI) is available at the transmitter. However, a more realistic scenario is the one where only an estimation of the channel is available at the transmitter. The relation between the true channel $\mathbf{H}$ and the estimated channel $\hat{\mathbf{H}}$ is usually modeled as

$$\mathbf{H} = \hat{\mathbf{H}} + \mathbf{E} \tag{19}$$

where the matrix $\mathbf{E}$ represents the channel estimation error, and it is independent from $\hat{\mathbf{H}}$. The knowledge of $\hat{\mathbf{H}}$ is used by the transmitter to obtain the RCI precoding matrix. The entries of $\hat{\mathbf{H}}$ and $\mathbf{E}$ are i.i.d. complex Gaussian random variables with zero mean and variances $1 - \tau^2$ and τ^2, respectively. The value of $\tau \in [0, 1]$ depends on the quality and technique used for channel estimation. When $\tau = 0$ the CSI is perfectly known, whereas $\tau = 1$ corresponds to the case when no CSI is available at all.

Future research could analyze the performance of linear precoding in the presence of imperfect CSI, deriving the achievable secrecy sum-rate as a function of the channel estimation error variance τ^2. This would allow to study how the CSI estimation error must scale with the SNR, in order to maintain a given high-SNR rate gap to the case with perfect CSI, so that the multiplexing gain is not affected. More specifically, the case of frequency division duplex (FDD) systems could be studied. Assuming that users quantize their channel directions by using B bits and employing random vector quantization (RVQ), and that they feed the quantization index back to the transmitter [30, 31], it would be interesting to determine how many feedback bits are required by each user in order to maintain a constant gap to the case with perfect CSI.

6. Conclusions

Throughout this chapter, we presented an up-to-date summary of the research in the field of physical layer security for multiuser MIMO communications. Unlike classical cryptography, physical layer security does not require key distribution and management, it does not rely on the limited computational power of the eavesdroppers, and it does not employ complex encryption algorithms. For these reasons, it is suitable for large dynamic wireless networks, and it has been proposed to enhance the protection of confidential messages transmitted over wireless channels. In this chapter, we especially focused on the problem of secret communication in a multiuser MIMO system. We considered the general case where a multiantenna base station transmits independent confidential messages to a generic number of users. We assumed that the users can potentially act maliciously and eavesdrop on each other. For this system set-up, we presented some transmission schemes based on linear precoding. We discussed the performance of these schemes as well as the cost of simultaneously guaranteeing secrecy to multiple users.

It has been recently shown that, in the large SNR regime, a linear precoding scheme based on regularized channel inversion can achieve secrecy without reducing the sum-rate at no additional cost when the number of transmit antennas M is larger than the number of users K. If $K = M$, secrecy can be achieved with a small rate loss or, alternatively, without reducing the sum-rate at a cost of less than 4dB in terms of the power transmitted. However, the secrecy requirements limit the maximum number of users that can be served with a non-zero rate. When $K > M$, there is an optimal value of the SNR beyond which the achievable rate starts decreasing, and at large SNR the secrecy sum-rate achievable by RCI precoding is poor. The base station could prevent the secrecy sum-rate from decreasing by reducing the transmit power, and therefore the SNR, to the value that maximizes the secrecy sum-rate. This would result in a constant nonnegative high-SNR secrecy sum-rate. However, this strategy would not be effective if $K \geq 2M$.

Acknowledgements

The work of G. Geraci was supported in part by the Australian Government under International Postgraduate Research Scholarship, and in part by the Wireless Technologies Laboratory, CSIRO ICT Centre, Sydney, Australia. The work of J. Yuan was supported in part by the Australian Research Council Discovery Project (Grant DP120102607).

Author details

Giovanni Geraci[1,2,*] and Jinhong Yuan[1]

* Address all correspondence to: giovanni.geraci@yahoo.it

1 School of Electrical Eng. & Telecommunications, The University of New South Wales, Australia
2 Wireless and Networking Technologies Laboratory, CSIRO ICT Centre, Sydney, Australia

References

[1] Y. Liang, H. V. Poor, and S. Shamai (Shitz). *Information Theoretic Security*. Dordrecht, The Netherlands: Now Publisher, 2009.

[2] R. Liu and W. Trappe. Eds, *Securing Wireless Communications at the Physical Layer*. New York: Springer Verlag, 2010.

[3] A. Mukherjee, S. A. A. Fakoorian, J. Huang, and A. L. Swindlehurst. Principles of physical-layer security in multiuser wireless networks: Survey. 2010. arXiv:1011.3754.

[4] A. D. Wyner. The wire-tap channel. *Bell System Tech. J.*, 54:1355–1387, 1975.

[5] I. Csiszár and J. Körner. Broadcast channels with confidential messages. *IEEE Trans. Inf. Theory*, 24(3):339–348, May 1978.

[6] S. Leung-Yan-Cheong and M. Hellman. The Gaussian wire-tap channel. *IEEE Trans. Inf. Theory*, 24(4):451–456, July 1978.

[7] J. Barros and M.R.D. Rodrigues. Secrecy capacity of wireless channels. In *Proc. IEEE Int. Symp. on Inform. Theory (ISIT)*, pages 356–360, july 2006.

[8] Z. Li, W. Trappe, and R. Yates. Secret communication via multi-antenna transmission. In *Proc. CISS*, March 2007.

[9] A. Khisti, G. Wornell, A. Wiesel, and Y. Eldar. On the Gaussian MIMO wiretap channel. In *Proc. IEEE Int. Symp. on Inform. Theory (ISIT)*, pages 2471–2475, 2007.

[10] A. Khisti and G.W. Wornell. Secure transmission with multiple antennas I: The MISOME wiretap channel. *IEEE Trans. Inf. Theory*, 56(7):3088–3104, July 2010.

[11] S. Goel and R. Negi. Guaranteeing secrecy using artificial noise. *IEEE Trans. Wireless Commun.*, 7(6):2180–2189, 2008.

[12] X. Zhou and M.R. McKay. Secure transmission with artificial noise over fading channels: Achievable rate and optimal power allocation. *IEEE Trans. Veh. Technol.*, 59(8):3831–3842, Oct. 2010.

[13] A. Mukherjee and A.L. Swindlehurst. Robust beamforming for security in MIMO wiretap channels with imperfect CSI. *IEEE Trans. Signal Process.*, 59(1):351–361, Jan. 2011.

[14] E. Ekrem and S. Ulukus. The secrecy capacity region of the Gaussian MIMO multi-receiver wiretap channel. *IEEE Trans. Inf. Theory*, 57(4):2083–2114, April 2011.

[15] R. Liu and H.V. Poor. Secrecy capacity region of a multiple-antenna Gaussian broadcast channel with confidential messages. *IEEE Trans. Inf. Theory*, 55(3):1235–1249, 2009.

[16] R. Liu, T. Liu, H.V. Poor, and S. Shamai. Multiple-input multiple-output Gaussian broadcast channels with confidential messages. *IEEE Trans. Inf. Theory*, 56(9):4215–4227, 2010.

[17] G. Geraci, J. Yuan, A. Razi, and I. B. Collings. Secrecy sum-rates for multi-user MIMO linear precoding. In *Proc. IEEE Int. Symp. on Wireless Commun. Systems (ISWCS)*, Nov. 2011.

[18] G. Geraci, M. Egan, J. Yuan, A. Razi, and I. B. Collings. Secrecy sum-rates for multi-user MIMO regularized channel inversion precoding. *IEEE Trans. Commun.*, 2012. to appear. Available: http://arxiv.org/abs/1207.5063.

[19] G. Geraci, J. Yuan, and I. B. Collings. Large system analysis of the secrecy sum-rates with regularized channel inversion precoding. In *Proc. IEEE Wireless Commun. Networking Conference (WCNC)*, Apr. 2012.

[20] Q.H. Spencer, C.B. Peel, A.L. Swindlehurst, and M. Haardt. An introduction to the multi-user MIMO downlink. *IEEE Comms. Mag.*, 42(10):60–67, Oct. 2004.

[21] Qinghua Li, Guangjie Li, Wookbong Lee, Moon Lee, D. Mazzarese, B. Clerckx, and Zexian Li. MIMO techniques in WiMAX and LTE: a feature overview. *IEEE Comms. Mag.*, 48(5):86–92, May 2010.

[22] Q.H. Spencer, A.L. Swindlehurst, and M. Haardt. Zero-forcing methods for downlink spatial multiplexing in multiuser MIMO channels. *IEEE Trans. Signal Process.*, 52(2):461–471, 2004.

[23] T. Yoo and A. Goldsmith. On the optimality of multiantenna broadcast scheduling using zero-forcing beamforming. *IEEE J. Sel. Areas Commun.*, 24(3):528–541, March 2006.

[24] C. B. Peel, B. M. Hochwald, and A. L. Swindlehurst. A vector-perturbation technique for near-capacity multiantenna multiuser communication - Part I: Channel inversion and regularization. *IEEE Trans. Commun.*, 53(1):195–202, January 2005.

[25] M. Joham, W. Utschick, and J.A. Nossek. Linear transmit processing in MIMO communications systems. *IEEE Trans. Signal Process.*, 53(8):2700–2712, August 2005.

[26] L. Sun and M.R. McKay. Eigen-based transceivers for the MIMO broadcast channel with semi-orthogonal user selection. *IEEE Trans. Signal Process.*, 58(10):5246–5261, Oct. 2010.

[27] S. Jin, M. R. McKay, X. Gao, and I. B. Collings. MIMO multichannel beamforming: SER and outage using new eigenvalue distributions of complex noncentral wishart matrices. *IEEE Trans. Commun.*, 56(3):424–434, 2008.

[28] V.K. Nguyen and J.S. Evans. Multiuser transmit beamforming via regularized channel inversion: A large system analysis. In *Proc. IEEE Global Commun. Conf. (GLOBECOM)*, pages 1–4, Dec. 2008.

[29] V. K. Nguyen, R. Muharar, and J. S. Evans. Multiuser transmit beamforming via regularized channel inversion: A large system analysis. *Technical Report*, Nov. 2009. Available: http://cubinlab.ee.unimelb.edu.au/ rmuharar/doc/manuscriptrevRusdha22- 11.pdf.

[30] N. Jindal. MIMO broadcast channels with finite-rate feedback. *IEEE Trans. Inf. Theory*, 52(11):5045–5060, November 2006.

[31] D. Ryan, I. B. Collings, I. V. L. Clarkson, and R. W. Heath Jr. Performance of vector perturbation multiuser MIMO systems with limited feedback. *IEEE Trans. Commun.*, 57(9):2633–2644, 2008.

Section 7

User Selection in Multi-User MIMO Systems

Chapter 7

Lattice Reduction-Based User Selection in Multiuser MIMO Systems

Qiaoyu Li, Ying Li, Lin Bai and Jinho Choi

Additional information is available at the end of the chapter

http://dx.doi.org/10.5772/57130

1. Introduction

In wireless communications, the spectral efficiency can be improved by exploiting the space domain when antenna arrays are used. In particular, space-division multiple access (SDMA) [1–3] can be adopted with various beamforming techniques. If both the transmitter and the receiver are equipped with multiple antennas, the resulting channel becomes a multiple-input-multiple-output (MIMO) channel, which can provide a rich spatial diversity gain. In MIMO systems, it is often desirable to use the maximum likelihood (ML) detection to jointly detect received signals for optimal performance and full receive diversity. However, since the complexity of the ML detection exponentially grows with the number of transmit antennas, the ML detection approach becomes impractical for high-dimensional detection problems. To derive low-complexity suboptimal MIMO detectors, various approaches based on the properties of lattice are considered. For example, using the Lenstra-Lenstra-Lovász (LLL) algorithm in [4], the lattice reduction (LR)-based low-complexity detectors are proposed in [5–8], which can provide a full receive diversity gain with a near-ML performance. The basic idea of the LR-based MIMO detection is to generate a nearly orthogonal basis for a given channel matrix to mitigate the effect of (multiple antenna) interference.

Due to users' different locations and channel conditions, it is possible to exploit another diversity gain in a multiuser system, where the throughput can be maximized by choosing the user of the strongest channel gain at a time. The resulting diversity gain is called the multiuser diversity gain [9]. Multiuser systems can be extended to the case of MIMO systems [10], where the multiuser MIMO user selection plays a key role in increasing the throughput of downlink channels [11]. It is noteworthy that, by viewing the multiuser MIMO system as virtual antennas in a single-user MIMO system, various antenna selection techniques can be applied to user selection [12, 13]. A mutual information-based criterion is proposed in [12] to select the antenna subset that maximizes the mutual information. In [13], a geometry-based criterion is developed with an LR-based linear detector to minimum the error probability. In general, user selection problems are combinatorial problems, and the complexity required to solve the problems could be prohibitively high for a large multiuser MIMO system. Thus, low-complexity suboptimal selection strategies are considered in [14–21], at the expense of degraded performance. In [14–17], a single antenna is selected at a time to maximize the throughput based on greedy selection schemes.

Although the achievable rate or related signal-to-noise ratio (SNR) can be used for the user selection criterion, it would be more practical to use a certain performance measure that is directly related to the performance of the actual detector or decoder employed. Therefore, it is desirable to derive a user selection criterion that can maximize the performance of the MIMO detector that is *actually* employed in a multiuser MIMO system.

In this chapter, for the user selection in uplink channels of a cellular system, where a single user is selected to transmit signals to a base station (BS) at a time, the error probability is used for the user selection criterion to choose the user with the smallest error probability for given MIMO detectors. Various user selection criteria will be derived with the ML detector, LR-based detectors and other low-complexity suboptimal detectors. It will be shown that a near-optimal performance with a full diversity gain (i.e., multiuser diversity and multiple antenna diversity) can be achieved using the proposed user selection criteria in this chapter with LR-based detectors.

Based on the single user selection criteria derived, we will extend them to support multiple users at a time. This extension of the user selection (i.e., multiple user selection) is not straightforward, because the multiple-user selection problem becomes a combinatorial problem. If an exhaustive search is used for multiple user selection when an LR-based MIMO detector is employed, LR needs to be performed for all the possible channel matrices composited by a group of subchannel matrices of the selected users. Unfortunately, this results in a high computational complexity, because the number of user combinations is large. Therefore, we will propose a greedy user selection algorithm to reduce the computational complexity at the expense of degraded performance when LR-based detectors are used. Moreover, to further reduce the computational complexity, an iterative LR updating algorithm will be investigated. Based on a theoretical analysis in this chapter, we can show that, with the combinatorial user selection, the LR-based detection can achieve the same diversity as the ML detector. Through simulations, we will compare the performance obtained by our selection criteria (i.e., combinatorial and greedy ones) to other existing approaches.

With the LR-based detection employed, simulation results will confirm that our combinatorial user selection can provide the best performance, whereas the performance of the greedy user selection scheme could approach that of the combinatorial approach as the correlation between possible composite channel matrices decreases. It will also be shown that our greedy user selection provides a better performance and a significantly reduced complexity compared with other approaches.

2. System model

In this section, we introduce the model of multiuser MIMO system together with several MIMO detection techniques.

2.1. Multiuser MIMO system

Consider the multiuser MIMO system with K users in uplink channels, where each user is equipped with P transmit antennas, and the base station (BS) is equipped with N receive antennas. Each user has an $N \times P$ channel matrix and a $P \times L$ signal matrix, which are denoted by $\mathbf{H}_k$ and $\mathbf{S}_k$, respectively, where $k \in \{1, 2, ..., K\}$. Here, L is the number of symbols transmitted by a user. It is assumed that all the users share a common uplink channel and that M users can access the channel at a time, where $M = \lfloor N/P \rfloor$. The channel is assumed to be a quasi-static block fading channel, with its channel matrix not varying over a time slot duration of L symbols. Here, a set of the M users who can access the channel could be updated for every time slot interval. Note that this selection problem can also be regarded as that with virtual antennas in a single-user MIMO system, where MP antennas are selected out of KP available antennas. Let $k_{(m)}$ be the mth selected user's index. For convenience, define the set of the selected users' indexes as $\mathcal{K} = \left\{k_{(1)}, k_{(2)}, ..., k_{(M)}\right\}$. Then, over a slot duration, the received signal at the BS is given by

$$\mathbf{Y}_{\mathcal{K}} = \mathbf{H}_{\mathcal{K}} \mathbf{S}_{\mathcal{K}} + \mathbf{N}, \tag{1}$$

where $\mathbf{H}_{\mathcal{K}}$, $\mathbf{S}_{\mathcal{K}}$, and $\mathbf{N}$ are the $N \times MP$ composite channel matrix, the $MP \times L$ transmitted signal matrix, and the $N \times L$ background noise matrix, respectively. We assume that each column vector of $\mathbf{N}$ is an independent zero-mean circularly symmetric complex Gaussian (CSCG) random vector with $\mathbb{E}\left\{\mathbf{n}_l \mathbf{n}_l^{\mathrm{H}}\right\} = N_0 \mathbf{I}$, where $\mathbf{n}_l$ denotes the lth column of $\mathbf{N}$. Note that $\mathbf{H}_{\mathcal{K}} = \left[\mathbf{H}_{k_{(1)}}, ..., \mathbf{H}_{k_{(M)}}\right]$ and that $\mathbf{S}_{\mathcal{K}} = \left[\mathbf{S}_{k_{(1)}}, ..., \mathbf{S}_{k_{(M)}}\right]$.

Throughout this chapter, we assume that the channel state information (CSI) is perfectly known at the receiver. Furthermore, the following assumptions are used to derive user selection methods.

A1) The elements of $\mathbf{S}_{\mathcal{K}}$ have a common signal alphabet, denoted by $\mathcal{S}$, and $\mathcal{S} \subset \mathbb{Z} + j\mathbb{Z}$, where $\mathbb{Z}$ denotes the set of integer numbers and $j = \sqrt{-1}$. Furthermore, let $\mathcal{S}^A$ represent the A-dimensional Cartesian product of $\mathcal{S}$.

A2) The transmitted signals are uncoded. This implies that the user selection criteria in this chapter are based on uncoded bit error rate (BER). For uncoded signals, we can assume $L = 1$ (Note that this assumption is used to simplify the derivation of user selection criteria, while the length of slot can be any number). Thus, $\mathbf{Y}_{\mathcal{K}}$, $\mathbf{S}_{\mathcal{K}}$, and $\mathbf{N}$ are vectors and will be denoted by $\mathbf{y}_{\mathcal{K}}$, $\mathbf{s}_{\mathcal{K}}$, and $\mathbf{n}$, respectively.

2.2. MIMO detection

MIMO detection plays an important role in MIMO receivers. Within this chapter, several well known MIMO detectors including the ML detector, linear detectors, and successive interference cancellation (SIC) detectors, together with LR are considered.

2.2.1. ML and linear detection

For the sake of convenience, we omit the index set $\mathcal{K}$. The ML detection is given by

$$\hat{\mathbf{s}}_{\text{ml}} = \arg \min_{\mathbf{s} \in \mathcal{S}^{MP}} \|\mathbf{y} - \mathbf{H}\mathbf{s}\|^2, \quad (2)$$

where the complexity grows exponentially with MP.

Alternatively, an estimate of $\mathbf{s}$ can be obtained by a linear transformation as follows:

$$\hat{\mathbf{s}} = \mathbf{W}\mathbf{y}, \quad (3)$$

where $\mathbf{W}$ is a linear filter that is given by $\mathbf{W} = \left(\mathbf{H}^{\text{H}}\mathbf{H} + c\mathbf{I}\right)^{-1}\mathbf{H}$. If $c = 0$, the linear detector corresponds to the zero-forcing (ZF) detector, while the minimum mean square error (MMSE) detector is obtained if $c = N_0/E_s$. Here, E_s is the symbol energy and it is assumed that $\mathbb{E}\left\{\mathbf{s}\mathbf{s}^{\text{H}}\right\} = E_s\mathbf{I}$.

To improve the performance of the detector, the LR is performed in the LR-based detection. A complex valued matrix can be converted into a real valued matrix for the LR as in [7]. Alternatively, the LR can be directly performed with a complex valued matrix as in [6], [8]. For convenience, in this chapter, we assume that the LR is performed with complex valued matrices.

For a given channel matrix $\mathbf{H}$, the LR basis can be found as follows:

$$\mathbf{H} = \mathbf{G}\mathbf{U}, \quad (4)$$

where $\mathbf{U}$ is an (complex) integer unimodular matrix and $\mathbf{G}$ is a matrix whose column vectors are nearly orthogonal. The received signal can be rewritten as

$$\mathbf{y} = \mathbf{H}\mathbf{s} + \mathbf{n} = \mathbf{G}\mathbf{c} + \mathbf{n}, \quad (5)$$

Under the MMSE criteria, the linear filter of LR-based MMSE linear detector is given by $\mathbf{W} = \left(\mathbf{G}^{\text{H}}\mathbf{G} + \frac{N_0}{E_s}\mathbf{U}^{\text{H}}\mathbf{U}\right)^{-1}\mathbf{G}^{\text{H}}$.

2.2.2. SIC detection

An SIC detector is not a linear detector due to its cancellation operation. In [7], the LR-based SIC detectors are proposed. To generalize the LR-based SIC detector, define the extended channel matrix as $\mathbf{H}_{\text{ex}} = [\mathbf{H}^{\text{T}} \ \sqrt{c}\mathbf{I}]^{\text{T}}$. The LR basis can be found as

$$\mathbf{H}_{\text{ex}} = \mathbf{G}_{\text{ex}}\mathbf{U}_{\text{ex}}, \quad (6)$$

where $\mathbf{U}_{\text{ex}}$ is a complex integer unimodular matrix and $\mathbf{G}_{\text{ex}}$ is a matrix whose column vectors are nearly orthogonal. If the LR basis is not used, $\mathbf{U}_{\text{ex}} = \mathbf{I}$ (i.e., $\mathbf{G}_{\text{ex}} = \mathbf{H}_{\text{ex}}$).

Note that the size of $\mathbf{H}_{\text{ex}}$ is the same as that of $\mathbf{G}_{\text{ex}}$ which is $2N \times MP$. Let the QR factorization of $\mathbf{G}_{\text{ex}}$ be $\mathbf{G}_{\text{ex}} = \mathbf{QR}$, where $\mathbf{Q}$ is a matrix whose column vectors are orthonormal and $\mathbf{R}$ is upper triangular. Let $\mathbf{y}_{\text{ex}} = [\mathbf{y}^{\text{T}}\ \mathbf{0}]^{\text{T}}$ and $\mathbf{n}_{\text{ex}} = [\mathbf{n}^{\text{T}}\ -\sqrt{c}\mathbf{s}^{\text{T}}]^{\text{T}}$. This results in $\mathbf{y}_{\text{ex}} = \mathbf{H}_{\text{ex}}\mathbf{s} + \mathbf{n}_{\text{ex}}$. Then, the LR-based SIC detection can be carried out with the following signal:

$$\mathbf{Q}^{\text{H}}\mathbf{y}_{\text{ex}} = \mathbf{Q}^{\text{H}}\mathbf{G}_{\text{ex}}\mathbf{U}_{\text{ex}}\mathbf{s} + \mathbf{Q}^{\text{H}}\mathbf{n}_{\text{ex}} = \mathbf{Rc} + \bar{\mathbf{n}}, \tag{7}$$

where $\mathbf{c} = \mathbf{U}_{\text{ex}}\mathbf{s}$ and $\bar{\mathbf{n}} = \mathbf{Q}_{\text{ex}}^{\text{H}}\mathbf{n}_{\text{ex}}$. Since the statistical properties of $\bar{\mathbf{n}}$ and $\mathbf{n}$ are the same, we will use $\mathbf{n}$ to denote $\bar{\mathbf{n}}$. Note that $\mathbf{n}$ also includes the self-interference as mentioned in [7].

The SIC detection can be carried out with (7). The elements of the last row, the MPth layer, are detected first. Then, their contributions in the second last row are canceled and the signals of the $(MP-1)$th row are detected. This operation is repeated up to the first row.

3. Single user selection criteria

In this section, we derive user selection criteria depending on the type of actually employed MIMO detector, where a single user is selected to transmit signals to a BS at a time. Suppose that user k is chosen, the system model in (1) is simplified as

$$\mathbf{Y}_k = \mathbf{H}_k\mathbf{S}_k + \mathbf{N}, \tag{8}$$

For detection method, the ML detector and two suboptimal detectors will be considered: one is the linear detector and the other is the SIC detector. As for the two suboptimal detectors, the LR is applied for better performance [6][7].

3.1. ML detector

Assuming that user k is selected, we omit the user index k for the sake of simplicity. To derive the selection criterion, we can consider the pairwise error probability (PEP). Suppose that $\mathbf{s}_{(1)}$ is transmitted, while $\mathbf{s}_{(2)}$ is erroneously detected. Then, the PEP is given by

$$\begin{aligned} P\left(\mathbf{s}_{(1)} \to \mathbf{s}_{(2)}\right) &= \Pr\left(\left\|\mathbf{y} - \mathbf{H}\mathbf{s}_{(2)}\right\|^2 \leq \left\|\mathbf{y} - \mathbf{H}\mathbf{s}_{(1)}\right\|^2\right) \\ &= \mathcal{Q}\left(\sqrt{\frac{\|\mathbf{H}\boldsymbol{\Delta}\|}{2N_0}}\right), \end{aligned} \tag{9}$$

where $\mathcal{Q}(x) = \int_x^{\infty} \frac{1}{\sqrt{2\pi}} e^{z^2/2} \mathrm{d}z$ and $\boldsymbol{\Delta} = \mathbf{s}_{(1)} - \mathbf{s}_{(2)}$. Then, the following upper bound can be obtained as

$$P\left(\mathbf{s}_{(1)} \to \mathbf{s}_{(2)}\right) \leq \mathcal{Q}\left(\sqrt{\frac{\|\mathbf{H}\bar{\mathbf{d}}\|^2}{2N_0}}\right), \tag{10}$$

where

$$\bar{\mathbf{d}} = \arg \min_{\mathbf{d} \in \mathcal{D}, \mathbf{d} \neq \mathbf{0}} ||\mathbf{H}\mathbf{d}||^2. \quad (11)$$

Here, $\mathcal{D} = \left\{ \mathbf{d} = \mathbf{s} - \mathbf{s}' \middle| \mathbf{s}, \mathbf{s}' \in \mathcal{S}^P \right\} \subset \mathbb{Z}^P + j\mathbb{Z}^P$. For convenience, denote by $S(\mathbf{H})$ the length of the shortest non-zero vector of the lattice generated by $\mathbf{H}$. Then, we can see that $S(\mathbf{H}) = ||\mathbf{H}\bar{\mathbf{d}}||$. From (10), if the ML detector is employed, the user selection criterion to minimize the error probability becomes

$$k^* = \arg \max_k S(\mathbf{H}_k). \quad (12)$$

Throughout this chapter, the user selection criterion in (12) is referred to as the max-min distance (MDist) criterion as $S(\mathbf{H})$ is the minimum distance of the lattice generated by $\mathbf{H}$.

The problem to find a non-zero shortest vector in a lattice is called the shortest vector problem (SVP) and known to be NP-hard. For an approximation, the LLL algorithm in [4], which has a polynomial time complexity, can be used.

Another approximation can be considered by relaxing the constraint on $\boldsymbol{\Delta}$. We have

$$||\mathbf{H}\boldsymbol{\Delta}||^2 = \boldsymbol{\Delta}^{\mathrm{H}}\mathbf{H}^{\mathrm{H}}\mathbf{H}\boldsymbol{\Delta} \geq ||\boldsymbol{\Delta}||^2 \lambda_{\min}\left(\mathbf{H}^{\mathrm{H}}\mathbf{H}\right), \quad (13)$$

where $\lambda_{\min}(\mathbf{A})$ stands for the minimum eigenvalue of $\mathbf{A}$. This shows that the selection criterion can be based on the minimum eigenvalue of the channel matrix, i.e.,

$$k^* = \arg \max_k \lambda_{\min}\left(\mathbf{H}_k^{\mathrm{H}}\mathbf{H}_k\right). \quad (14)$$

Thus, each user can feed back its minimum eigenvalue of the channel matrix and the user who has the maximum $\lambda_{\min}\left(\mathbf{H}_k^{\mathrm{H}}\mathbf{H}_k\right)$ can be selected to access the channel. This selection criterion is referred to as the max-min eigenvalue (ME) criterion throughout this chapter.

3.2. Linear detectors

As the SNR increases, we have $c \to 0$ (in this case, the MMSE detector becomes the ZF detector) and the PEP has the following upper bound:

$$\begin{aligned} P\left(\mathbf{s}_{(1)} \to \mathbf{s}_{(2)}\right) &= \mathcal{Q}\left(\frac{||\boldsymbol{\Delta}||^2}{\sqrt{2N_0\boldsymbol{\Delta}^{\mathrm{H}}(\mathbf{H}^{\mathrm{H}}\mathbf{H})^{-1}\boldsymbol{\Delta}}}\right) \\ &\leq \mathcal{Q}\left(\sqrt{\frac{\lambda_{\min}(\mathbf{H}^{\mathrm{H}}\mathbf{H})}{2N_0}||\boldsymbol{\Delta}||^2}\right), \end{aligned} \quad (15)$$

because $\mathcal{Q}(\cdot)$ is a decreasing function and $\boldsymbol{\Delta}^{\mathrm{H}}(\mathbf{H}^{\mathrm{H}}\mathbf{H})^{-1}\boldsymbol{\Delta} \le \lambda_{\max}(\mathbf{H}^{\mathrm{H}}\mathbf{H})^{-1}\|\boldsymbol{\Delta}\|^2 = \frac{\|\boldsymbol{\Delta}\|^2}{\lambda_{\min}(\mathbf{H}^{\mathrm{H}}\mathbf{H})}$. Therefore, the ME criterion in (14) can be used for the user selection criterion.

It is important to note that this ME criterion is valid for the LR-based linear detectors [6], [7]. Let $\mathbf{c}_{(i)} = \mathbf{U}\mathbf{s}_{(i)}$, $i = 1, 2$. Then, from (15), the PEP is bounded as

$$P\left(\mathbf{s}_{(1)} \to \mathbf{s}_{(2)}\right) \le \mathcal{Q}\left(\sqrt{\frac{\lambda_{\min}\left(\mathbf{G}^{\mathrm{H}}\mathbf{G}\right)\|\boldsymbol{\Delta}_{\mathbf{U}}\|^2}{2N_0}}\right), \tag{16}$$

where $\boldsymbol{\Delta}_{\mathbf{U}} = \mathbf{c}_{(1)} - \mathbf{c}_{(2)} = \mathbf{U}\left(\mathbf{s}_{(1)} - \mathbf{s}_{(2)}\right)$. From (16), the selection criterion becomes

$$k^* = \arg\max_k \lambda_{\min}\left(\mathbf{G}_k^{\mathrm{H}}\mathbf{G}_k\right), \tag{17}$$

where $\mathbf{G}_k$ is the reduced basis from $\mathbf{H}_k$. This ME criterion is the same as that in (14) except that the channel matrix $\mathbf{H}_k$ is replaced by its reduced one $\mathbf{G}_k$.

3.3. SIC detectors

As the LR is performed, the column vectors of $\mathbf{G}_{\mathrm{ex}}$ would be nearly orthogonal. In other words, the upper off-diagonal elements of $\mathbf{R}$ would be small. Thus the SIC detection performance would mainly depend on the diagonal elements of $\mathbf{R}$. For convenience, let $c = 0$ (this is the case when $N_0 \to 0$ or high SNR). Let $r_{p,p}^{(k)}$ denote the (p, p)th element of $\mathbf{R}$ from the kth user's channel $\mathbf{H}_k$. Then, ignoring the interference terms (as they are canceled when the detection of the lower layers is successfully carried out with no error), the SNR of the pth layer of $\mathbf{H}_k$ becomes $\gamma_p^{(k)} = \frac{|r_{p,p}^{(k)}|^2}{N_0}$. From this, the selection criterion can be given by

$$k^* = \arg\max_k \left\{\min_p \left|r_{p,p}^{(k)}\right|\right\}. \tag{18}$$

This selection criterion is referred to as the max-min diagonal term (MD) criterion.

The MD criterion is also closely related to the minimum error probability criterion when the SNR is high. For convenience, let $\mathbf{x} = \mathbf{Q}^{\mathrm{H}}\mathbf{y}$. Then, (18) is rewritten as

$$\mathbf{x} = \mathbf{R}\mathbf{c} + \mathbf{n}. \tag{19}$$

Let n_p denote the pth element of $\mathbf{n}$. Then, the LR-based SIC detection at the Pth layer does not have error if $\frac{|n_P|}{|r_{P,P}|} < \frac{1}{2}$ or $|n_P|^2 < \frac{|r_{P,P}|^2}{4}$. Thus, the LR-based SIC detection would have no error across all the layers if $|n_p|^2 < \frac{|r_{p,p}|^2}{4}$, for all p. The probability of no error can be lower bounded as

$$\begin{aligned}\Pr(\text{no error}) &\geq \Pr\left(|n_p|^2 < \frac{|r_{p,1}|^2}{4}, \forall p\right) \\ &= \prod_{p=1}^{P} \Pr\left(|n_p|^2 < \frac{|r_{p,p}|^2}{4}\right).\end{aligned} \tag{20}$$

Since $|n_p|^2$ is a chi-square random variable with 2 degrees of freedom (or an exponential random variable), we have $\Pr\left(|n_p|^2 < \frac{|r_{p,p}|^2}{4}\right) = 1 - \exp\left(-\frac{|r_{p,p}|^2}{4N_0}\right)$. Thus, from (20), the probability of error can be given by

$$\begin{aligned}\Pr(\text{error}) &\leq 1 - \prod_{p=1}^{Q}\left(1 - \exp\left(\frac{|r_{p,p}|^2}{4N_0}\right)\right) \\ &\simeq \exp\left(-\min_p \frac{|r_{p,p}|^2}{4N_0}\right) \quad \text{as } N_0 \to 0.\end{aligned} \tag{21}$$

Therefore, to minimize the probability of error, the user who has the maximum $\min_p |r_{p,p}|$ can be selected.

4. Selection criteria with multiple users

To maximize the performance, if $M = 1$, the user who can have the minimum PEP is chosen for a given MIMO detector. In Section 3, a few user selection criteria are derived, depending on the types of actually employed MIMO detectors. Note that only one user is selected (i.e., $M = 1$) in Section 3. To extend the user selection criteria to the case of $M > 1$ here and in the consecutive sections, we consider the combinatorial and greedy user selection criteria.

4.1. ML and MMSE selection criteria

For a given $M > 1$, the set of the users who can access the channel can be found using the MDist or ME user selection criterion as follows:

$$\mathcal{K}_{\text{MDist}} = \arg\max_{\mathcal{K}} \mathcal{S}\left(\mathbf{H}_{\mathcal{K}}\right) \tag{22}$$

or

$$\mathcal{K}_{\text{ME}} = \arg\max_{\mathcal{K}} \lambda_{\min}\left(\mathbf{H}_{\mathcal{K}}^{\text{H}}\mathbf{H}_{\mathcal{K}}\right) \tag{23}$$

respectively. If the ML detector is employed, the MDist user selection criterion can be used to choose the M users who can have the lowest BER, whereas the ME criterion is used to choose the M users with the highest worst SNR (i.e., max-min SNR).

4.2. LR-based MMSE and MMSE-SIC selection criteria

In this subsection, the user selection criteria with LR-based detectors in Section 3 are extended to the case of $M > 1$, where the number of transmit layers are extended to MP, compared to P in the case of $M = 1$.

The MD criterion derived in Section 3, with $M = 1$ for the LR-based MMSE-SIC detection, can be extended to the case with $M > 1$ as follows:

$$\mathcal{K}_{\text{MD}} = \arg\max_{\mathcal{K}} \left\{ \min_{q} \left| r_{q,q}^{(\mathcal{K})} \right| \right\} \tag{24}$$

and the ME criterion for the LR-based MMSE detection can also be modified as

$$\mathcal{K}_{\text{ME}} = \arg\max_{\mathcal{K}} \lambda_{\min} \left(\mathbf{G}_{\mathcal{K}}^{\text{H}} \mathbf{G}_{\mathcal{K}} \right). \tag{25}$$

The user selection based on (22), (23), (24), and (25) is called the combinatorial user selection, because the users have to be selected by combinatorial (or exhaustive) search.

5. LR-based greedy user selection using an updating method

The computational complexity of the user selection under the criteria derived in Section 4 grows rapidly with M or K as they are all combinatorial optimization problems. Thus, it is desirable to derive low complexity approaches for the user selection. In this section, we propose low complexity greedy approaches for the user selection. Note that we focus on the greedy user selection with a LR-based MIMO detector only as its performance is comparable to that of the ML detector and, more importantly, we can derive a computationally efficient LR updating method in conjunction with greedy user selection.

5.1. LR-based greedy user selection

The user selection approaches in Section 4 have the complexity that becomes prohibitively high as M or K increases, because there are $U = \prod_{i=0}^{M-1}(K-i)$ possible user index sets. For each user index set, an LR of an $N \times MP$ complex channel matrix is to be performed. For example, when $K = 10$, $M = N = 4$ and $P = 1$, $10 \times 9 \times 8 \times 7 = 5040$ LRs of 4×4 complex-valued channel matrices should be carried out.

To reduce the computational complexity in the user selection, we consider a greedy approach when a LR-based MIMO detector is employed. The resulting approach is called the LR-based greedy (LRG) user selection, which is of course suboptimal. The LRG user selection algorithm is summarized as follows:

1. Let $m = 1$ and $\bar{\mathcal{K}} = \{1, \ldots, K\}$. In order to select the first user, we can use any criterion. For example, if the ME criterion is used, we have

$$k_{(1)} = \arg\max_{k \in \bar{\mathcal{K}}} \lambda_{\min} \left(\mathbf{G}_k^{\text{H}} \mathbf{G}_k \right), \tag{26}$$

where $\mathbf{G}_k$ represents the LBR matrix of $\mathbf{H}_k$ or $\mathbf{H}_{\mathrm{ex},k} = \left[\mathbf{H}_k^{\mathrm{T}} \; \sqrt{\frac{N_0}{E_s}}\mathbf{I}_N\right]^{\mathrm{T}}$ (for the LR-based MMSE detector). Once the first user is chosen, we update $\bar{\mathcal{K}}$ as $\bar{\mathcal{K}} \Leftarrow \bar{\mathcal{K}} \setminus \left\{k_{(1)}\right\}$. In addition, we let $\mathbf{H}_{(1)} = \mathbf{H}_{k_{(1)}}$.

2. Let $m \Leftarrow m+1$ and $\mathbf{H}_{(m),k} = \left[\mathbf{H}_{(m-1)} \; \mathbf{H}_k\right]$, $k \in \bar{\mathcal{K}}$. The mth user can be chosen if the ME criterion is used as

$$k_{(m)} = \arg\max_{k \in \bar{\mathcal{K}}} \lambda_{\min}\left(\mathbf{G}_{(m),k}^{\mathrm{H}} \mathbf{G}_{(m),k}\right), \tag{27}$$

where $\mathbf{G}_{(m),k}$ is the LBR matrix of $\mathbf{H}_{(m),k}$ or $\mathbf{H}_{\mathrm{ex},(m),k} = \left[\mathbf{H}_{(m),k}^{\mathrm{T}} \; \sqrt{\frac{N_0}{E_s}}\mathbf{I}_N\right]^{\mathrm{T}}$. Once the mth user is found, we update as follows:

$$\begin{aligned} &\text{add } k_{(m)} \text{ to the index set of the selected users, } \mathcal{K}, \\ &\bar{\mathcal{K}} \Leftarrow \bar{\mathcal{K}} \setminus k_{(m)}, \\ &\mathbf{H}_{(m)} = \mathbf{H}_{(m),k_{(m)}}. \end{aligned} \tag{28}$$

3. If $m = M$, stop. Otherwise, go to 2).

Note that in this algorithm, the $N \times mP$ complex-valued matrix $\mathbf{H}_{(m)}$ denotes the channel matrix for the first m selected users, while the $N \times P$ complex-valued matrix $\mathbf{H}_{k_{(m)}}$ represents the channel matrix for the selected user in the mth selection with the index $k_{(m)}$, where $k_{(m)} \in \bar{\mathcal{K}}$ and $\bar{\mathcal{K}} = \{1, \ldots, K\} \setminus \left\{k_{(1)}, \ldots, k_{(m-1)}\right\}$.

In the LRG user selection, the number of required LR operations is $\sum_{i=1}^{M}(K-i+1)$ and the matrix size for LR in selecting the mth user is $N \times mP$. Using the upper bound on the average complexity of LR studied in [8], we can show that the complexity of LRG is upper-bounded as $\sum_{i=1}^{M}(K-i+1)O\left((iP)^3 N \log(iP)\right)$ (Note that when $P = 1$, no LR is required for the first user selection, where the complexity of LRG reduces to $\sum_{i=2}^{M}(K-i+1)O\left((iP)^3 N \log(iP)\right)$). On the other hand, the number of required LR operations in the combinatorial user selection according to (24) or (25) is $\prod_{i=1}^{M}(K-i+1)$ and the matrix size for LR is always $N \times MP$, which leads to its complexity that is upper-bounded as $\prod_{i=1}^{M}(K-i+1)O\left((MP)^3 N \log(MP)\right)$. This shows a significant computational complexity reduction. However, since the LRG user selection does not jointly select M users, there will be performance loss.

Note that the ME criterion is used in above for illustration purposes. The MD criterion can also be used for the LRG user selection with the LR-based MMSE-SIC detector.

5.2. A complexity efficient method for LR updating

We note that in the LRG user selection, the LR operation is repeatedly performed for each updated channel matrix. For instance, at the mth user selection, a LR is carried out with the complex-valued channel matrix $\mathbf{H}_{(m)} = \left[\mathbf{H}_{(m-1)} \; \mathbf{H}_k\right]$ as shown in (27), where $\mathbf{H}_k$ contains P newly added column vectors and the other $(m-1)P$ column vectors in $\mathbf{H}_{(m)}$ are already chosen and LBR. Instead of performing a new LR on all of the mP column vectors in $\mathbf{H}_{(m)}$, by utilizing the established $(m-1)P$ LBR vectors, we can derive a computationally efficient LR updating method with new P column vectors, which is referred to as the Updated Basis LR (UBLR) in this paper. The resulting user selection scheme is referred to as the UBLR-based greedy (UBLRG)[1] user selection.

The UBLR algorithm is based on the complex-LLL(CLLL) algorithm [8]. Suppose that LR is performed by the CLLL algorithm in order to transform a given basis (a complex-valued channel matrix $N \times mP$ $\mathbf{H}_{(m)}$) into a new $N \times mP$ basis $\mathbf{G}_{(m)}$ consisting of nearly orthogonal basis vectors. A basis $\mathbf{G}_{(m)}$ is called a reduced basis of a lattice with parameter δ if $\mathbf{G}_{(m)}$ is QR factorized as $\mathbf{G}_{(m)} = \mathbf{Q}_{(m)}\mathbf{R}_{(m)}$, where $\mathbf{Q}_{(m)}$ is unitary, $\mathbf{R}_{(m)}$ is upper triangular, and the elements of $\mathbf{R}_{(m)}$ satisfy the following inequalities:

$$| \Re(r_{\ell,\rho}) | \leq \frac{1}{2} | r_{\ell,\ell} | \quad \text{and} \quad | \Im(r_{\ell,\rho}) | \leq \frac{1}{2} | r_{\ell,\ell} | \quad \text{for } 1 \leq \ell < \rho \leq mP \tag{29}$$

and

$$\delta |r_{\rho-1,\rho-1}|^2 \leq |r_{\rho,\rho}|^2 + |r_{\rho-1,\rho}|^2 \quad \text{for } \rho = 2, \ldots, mP. \tag{30}$$

Here, $r_{p,q}$ denotes the (p,q)-th element of $\mathbf{R}_{(m)}$. The parameter δ is closely related to the quality of the reduced basis. In this paper, we assume $\delta = 3/4$[2] which is usually chosen for complexity and performance trade-off. For the initialization, let $\mathcal{A}'_{(m)} = \{\mathbf{Q}'_{(m)}, \mathbf{R}'_{(m)}, \mathbf{U}'_{(m)}\}$, where the QR factorization $\mathbf{H}_{(m)} = \mathbf{Q}'_{(m)}\mathbf{R}'_{(m)}$ and $\mathbf{U}'_{(m)} = \mathbf{I}_{mP}$. With $\{\mathbf{Q}_{(m)}, \mathbf{R}_{(m)}, \mathbf{U}_{(m)}\} = \{\mathbf{Q}'_{(m)}, \mathbf{R}'_{(m)}, \mathbf{U}'_{(m)}\}$ and $\rho = 2$, a version of CLLL algorithm is summarized as follows (note that since CLLL is used in UBLR, in Table 1, CLLL becomes part of UBLR).

a) To fulfill (29), a size-reduction is performed with the 1st to ρ^{th} columns of $\mathbf{R}_{(m)}$ and $\mathbf{U}_{(m)}$ (see rows (15)-(21) in Table 1).

b) As the basis of $\mathbf{R}_{(m)}$ is size-reduced according to (29), let $\rho \Leftarrow \rho + 1$ and go to step a) if (30) is fulfilled. Swap the $(\rho-1)^{\text{th}}$ and ρ^{th} columns in $\mathbf{R}_{(m)}$ and $\mathbf{U}_{(m)}$ if (14) is not satisfied and update $\{\mathbf{R}_{(m)}, \mathbf{Q}_{(m)}\}$. Let $\rho \Leftarrow \max(\rho - 1, 2)$ and go to step a) (see rows (22)-(32) in Table 1).

c) The algorithm is terminated if $\rho = mP$. The output of the CLLL reduced matrix $\mathbf{G}_{(m)}$ is given by the updated $\mathcal{A}_{(m)} = \{\mathbf{Q}_{(m)}, \mathbf{R}_{(m)}, \mathbf{U}_{(m)}\}$, i.e., $\mathbf{G}_{(m)} = \mathbf{Q}_{(m)}\mathbf{R}_{(m)} = \mathbf{H}_{(m)}\mathbf{U}_{(m)}$.

In our LRG user selection, at the mth user selection, the channel matrix of size $N \times P(m-1)$ (denoted by $\mathbf{H}_{(m-1)}$) is obtained from the previous user selections. Under the assumption that the CLLL has been performed with $\mathbf{H}_{(m-1)}$ and its CLLL reduced matrix $\mathbf{G}_{(m-1)}$ is

[1] Since the performance of the LRG and UBLRG user selection schemes are the same (in fact, UBLRG is a computationally efficient version of LRG), we now only consider UBLRG and assume that LRG and UBLRG are interchangeable.

[2] Here, δ is a factor selected to achieve a good quality-complexity trade-off [4]. We note that δ can be chose from $(\frac{1}{4}, 1)$ and $(\frac{1}{2}, 1)$ for the real and complex LLL algorithms, respectively.

INPUT: $\{\mathcal{A}_{(m-1)}, \mathcal{B}_{(m-1)}, \mathbf{H}_{(m-1)}, \mathbf{H}_{k_{(m)}}\}$.
OUTPUT: $\{\mathcal{A}_{(m)}, \mathcal{B}_{(m)}\}$.
$\mathbf{H}_{(m)} \leftarrow \left[\mathbf{H}_{(m-1)}\ \mathbf{H}_{k_{(m)}}\right]$
$\omega \leftarrow \text{size}(\mathbf{H}_{(m-1)}, 2)$
$\zeta \leftarrow \text{size}(\mathbf{H}_{(m)}, 2)$
$[\mathbf{Q}_{(m)}\ \mathbf{R}_{(m)}] \leftarrow \text{qr}(\mathbf{H}_{(m)})$
$\mathbf{U}_{(m)} \leftarrow \mathbf{I}_{\zeta}$
$\mathbf{U}_{(m)}(1:\omega, 1:\omega) \leftarrow \mathbf{U}_{(m-1)}$
$\mathbf{Q}_{(m)} \leftarrow \mathbf{Q}_{(m-1)}$
$\mathbf{R}_{(m)}(:, 1:\omega) \leftarrow \mathbf{R}_{(m-1)}$
for $\ell = 1 : \eta_{(m-1)}$
$\mathbf{R}_{(m)}\left(\mathbf{fl}_{(m-1,\ell)} - 1 : \mathbf{fl}_{(m-1,\ell)}, \omega + 1 : \zeta\right)$
$\leftarrow \mathbf{\Theta}_{(m-1,\ell)} \mathbf{R}_{(m)}\left(\mathbf{fl}_{(m-1,\ell)} - 1 : \mathbf{fl}_{(m-1,\ell)}, \omega + 1 : \zeta\right)$
end for
$\rho \leftarrow \omega + 1$
$\eta_{(m)} \leftarrow 0$
while $\rho \leq \zeta$
for $\ell = 1 : \rho - 1$
$\mu \leftarrow \lceil \mathbf{R}_{(m)}(\rho - \ell, \rho) / \mathbf{R}_{(m)}(\rho - \ell, \rho - \ell) \rfloor$
if $\mu \neq 0$
$\mathbf{R}_{(m)}(1 : \rho - \ell, \rho) \leftarrow \mathbf{R}_{(m)}(1 : \rho - \ell, \rho) - \mu \mathbf{R}_{(m)}(1 : \rho - \ell, \rho - \ell)$
$\mathbf{U}_{(m)}(:, \rho) \leftarrow \mathbf{U}_{(m)}(:, \rho) - \mu \mathbf{U}_{(m)}(:, \rho - \ell)$
end if
end for
if $\delta |(\mathbf{R}_{(m)}(\rho - 1, \rho - 1))|^2 > |\mathbf{R}_{(m)}(\rho, \rho)|^2 + |\mathbf{R}_{(m)}(\rho - 1, \rho)|^2$
$\eta_{(m)} \leftarrow \eta_{(m)} + 1$
Swap the $(\rho - 1)$-th and ρth columns in $\mathbf{R}_{(m)}$ and $\mathbf{U}_{(m)}$
$\mathbf{\Theta}_{(m,\eta_{(m)})} = \begin{bmatrix} \alpha^* & \beta \\ -\beta & \alpha \end{bmatrix}$ with $\alpha = \frac{\mathbf{R}_{(m)}(\rho-1,\rho-1)}{\|\mathbf{R}_{(m)}(\rho-1:\rho,\rho-1)\|}$, $\beta = \frac{\mathbf{R}_{(m)}(\rho,\rho-1)}{\|\mathbf{R}_{(m)}(\rho-1:\rho,\rho-1)\|}$
$\mathbf{fl}_{(m,\eta_{(m)})} \leftarrow \rho$
$\mathbf{R}_{(m)}(\rho - 1 : \rho, \rho - 1 : \zeta) \leftarrow \mathbf{\Theta}_{(m,\eta_{(m)})} \mathbf{R}_{(m)}(\rho - 1 : \rho, \rho - 1 : \zeta)$
$\mathbf{Q}_{(m)}(:, \rho - 1 : \rho) \leftarrow \mathbf{Q}_{(m)}(:, \rho - 1 : \rho) \mathbf{\Theta}^{\mathrm{T}}_{(m,\eta_{(m)})}$
$\rho \leftarrow \max\{\rho - 1, 2\}$
else
$\rho \leftarrow \rho + 1$
end if
end while

Table 1. The UBLR (based on the CLLL) algorithm at the mth user selection

available, we have $\mathbf{H}_{(m)} = \left[\mathbf{H}_{(m-1)}\ \mathbf{H}_{k_{(m)}}\right]$ which is the channel matrix for the first m selected users.

The UBLR algorithm is carried out to transform $\mathbf{H}_{(m)}$ into a reduced basis $\mathbf{G}_{(m)}$ by utilizing a given set of already available matrices $\mathcal{A}_{(m-1)} = \{\mathbf{Q}_{(m-1)}, \mathbf{R}_{(m-1)}, \mathbf{U}_{(m-1)}\}$ associated with the CLLL reduced matrix $\mathbf{G}_{(m-1)}$ in the previous $m - 1$ users selection, where $\mathbf{G}_{(m-1)} = \mathbf{Q}_{(m-1)}\mathbf{R}_{(m-1)} = \mathbf{H}_{(m-1)}\mathbf{U}_{(m-1)}$. The unimodular matrix $\mathbf{U}_{(m-1)}$ is employed to represent the column swaps in the CLLL, while $\mathbf{R}_{(m-1)}$ satisfies (29) and (30). The transformation algorithm for generating $\mathbf{G}_{(m)}$ in UBLR is summarized as follows.

Instead of starting the size-reduction of $\mathbf{R}'_{(m)}$ with the first two columns (the 1st to ρth columns, where $\rho = 2$ in a)), UBLR reduces the iteration by starting the size-reduction with $\rho = (m-1)P+1$. In this case, the iteration of size-reduction from that with $\rho = 2$ to that with $\rho = (m-1)P+1$ need to be obtained by updating $\mathcal{A}'_{(m)}$ from $\mathcal{A}_{(m-1)}$.

Since $\mathbf{R}'_{(m-1)}$ of size $N \times P(m-1)$ and $\mathbf{R}'_{(m)}$ of size $N \times Pm$ are upper triangular, it is straightforward to obtain that $\mathbf{R}'_{(m-1)} = \mathbf{R}'_{(m)}(:, 1:P(m-1))$, which results in that the size reduction and column swapping performed on the first $P(m-1)$ columns of $\mathbf{R}'_{(m)}$ are the same as those on $\mathbf{R}'_{(m-1)}$. Using $\mathbf{R}_{(m-1)}$, let $\mathcal{A}_{(m)} = \mathcal{A}'_{(m)}$ and $\mathbf{R}_{(m)}(:, 1:P(m-1)) = \mathbf{R}_{(m-1)}$. Then, we have the 1st to $P(m-1)$-th column vectors of $\mathbf{R}_{(m)}$ satisfying (29) and (30). From this, we can see that CLLL is partially performed on $\mathbf{R}_{(m)}$ by employing UBLR. Similarly, with $\mathbf{Q}_{(m)} = \mathbf{Q}_{(m-1)}$ and $\mathbf{U}_{(m)}(1:P(m-1), 1:P(m-1)) = \mathbf{U}_{(m-1)}$, $\{\mathbf{Q}_{(m)}, \mathbf{U}_{(m)}\}$ can be updated with low computational complexity from $\{\mathbf{Q}_{(m-1)}, \mathbf{U}_{(m-1)}\}$. Thus, from $\mathcal{A}_{(m-1)}$, UBLR is carried out to update the elements in $\mathcal{A}_{(m)}$ as shown in rows (6)-(8) in Table 1.

In addition, we note that, in row (8) of Table 1, we do not consider updating $\mathbf{R}_{(m)}(1:P(m-1), P(m-1)+1:Pm)$ in $\mathcal{A}_{(m)}$. It can be observed that when we perform a CLLL on $\mathbf{H}_{(m)}$ with the same operations of the CLLL for previous user selections, $\mathbf{R}_{(m)}(1:P(m-1), P(m-1)+1:Pm)$ will also be influenced. Hence, extra processing is necessary to recover $\mathbf{R}_{(m)}(1:P(m-1), P(m-1)+1:Pm)$ in $\mathcal{A}_{(m)}$. To this end, we define that $\mathcal{B}_{(m-1)} = \{\mathbf{\Theta}_{(m-1)}, \mathbf{fl}_{(m-1)}, \eta_{(m-1)}\}$, where $\mathbf{\Theta}_{(m-1)} = \{\mathbf{\Theta}_{(m-1,1)}, \cdots, \mathbf{\Theta}_{(m-1,\eta)}\}$, $\mathbf{fl}_{(m-1)} = \{\mathbf{fl}_{(m-1,1)}, \cdots, \mathbf{fl}_{(m-1,\eta)}\}$, and $\eta_{(m-1)} = \eta$. The operations of swapping and updating $\mathbf{R}_{(m-1)}$ and $\mathbf{Q}_{(m-1)}$ are kept in $\eta_{(m-1)}$, $\gamma_{(m-1)}$, and $\mathbf{\Theta}_{(m-1,\eta)}$, where $\eta_{(m-1)}$ keeps the number of swapping times, $\gamma_{(m-1)}$ keeps those columns involved in the swaps, and $\mathbf{\Theta}_{(m-1,\eta)}$ keeps the operations of column swaps. From the CLLL (see row (27) in Table 1), we note that $\mathbf{R}_{(m)}(1:P(m-1), P(m-1)+1:Pm)$ is generated by a transformation with $\mathbf{\Theta}_{(m)}$. Thus, using the information kept in $\mathcal{B}_{(m-1)}$, we can generate $\mathbf{R}_{(m)}(1:P(m-1), P(m-1)+1:Pm)$ as shown in rows (9)-(11) of Table 1.

With an updated $\mathcal{A}_{(m)}$, one CLLL can be carried out to generate the reduced basis $\mathbf{G}_{(m)}$. The calculation of this new basis generation starts with $\rho = (m-1)P+1$. Hence, the computational complexity of UBLR is evidently reduced as compared to employing one CLLL starting with $\rho = 2$. Note that since UBLR and CLLL generate the same LBR $\mathbf{G}_{(m)}$, they are expected to provide the same performance.

The UBLR algorithm of the mth user selection is summarized in Table 1. The inputs of the algorithm of the mth user selection are $\{\mathcal{A}_{(m-1)}, \mathcal{B}_{(m-1)}, \mathbf{H}_{(m-1)}, \mathbf{H}_{k_{(m)}}\}$, while the outputs are $\{\mathcal{A}_{(m)}, \mathcal{B}_{(m)}\}$. Note that for the first user selection, with its channel matrix $\mathbf{H}_{k_{(1)}}$ as the input, instead of using the UBLR, one CLLL is carried out to generate $\{\mathcal{A}_{(1)}, \mathcal{B}_{(1)}\}$ as the output. Since the outputs of the mth user selection are regarded as the inputs at the $(m+1)$-th user selection, the algorithm is recursively carried out from $m = 2$. The algorithm is terminated if $m = M$.

The complexity of CLLL and UBLR algorithms highly depends on the number of column swaps, which is denoted by the output parameter η. In Table 2, the average value of η per iteration is shown when the CLLL-based MMSE-SIC detector is used with the proposed LRG and UBLRG user selection. It is assumed that $K = 10$ and $N = 8$ for the two

	Average value of η							
Number of columns in $\mathbf{H}_{\mathcal{K}}$	2	3	4	5	6	7	8	Sum
LRG[1]	0.2909	0.9029	1.8022	3.0633	4.7711	7.2925	12.1228	30.2457
UBLRG[1]	0.2904	0.5851	0.8940	1.2708	1.7653	2.5620	4.7728	12.1404
LRG[2]	0.2926	n/a	1.7977	n/a	4.7663	n/a	12.0856	18.9422
UBLRG[2]	0.2879	n/a	1.4952	n/a	3.0191	n/a	7.3761	12.1783

Note that the superscript [1] denotes the case of $K = 10$, $N = 8$, $(M, P) = (8, 1)$ and the superscript [2] denotes the case of $K = 10$, $N = 8$, $(M, P) = (4, 2)$, respectively.

Table 2. The average value of η in the LRG and UBLRG user selection with the CLLL based MMSE-SIC detector is used.

possible cases of $(M, P) = (8, 1)$ and $(M, P) = (4, 2)$. Based on these results, we can observe that the complexity is significantly reduced if UBLR is employed. We also note that with the LRG, the complexity for the case of $(M, P) = (8, 1)$ is higher than that of $(M, P) = (4, 2)$ as expected (a large M implies a higher complexity). We can also show that the complexity of UBLRG is upper-bounded as $(K - M + 1)O\left((MP)^3 N \log(MP)\right) + \sum_{i=1}^{M-1} O\left((iP)^3 N \log(iP)\right)$. Compared to the complexity of LRG which is upper-bounded as $\sum_{i=1}^{M}(K - i + 1)O\left((iP)^3 N \log(iP)\right)$, the UBLRG scheme has a lower complexity, especially when large K and M are considered.

6. Diversity Analysis and Numerical Results

In this section, we consider the diversity gain of the combinatorial user selection approaches with various detectors, such as the ML, MMSE, and LR-based SIC detectors. We derive lower bounds on the diversity gain of them. Since the diversity gain analysis of the proposed greedy user selection approach is difficult, we rely on simulations, from which we can show that our proposed LRG/UBLRG user selection approach has a similar diversity gain and comparable performance to the combinatorial one. Throughout this section, we assume that the elements of the channel matrix $\mathbf{H}_{\mathcal{K}}$ are independent zero-mean CSCG random variables with variance σ_h^2.

6.1. Diversity Gain Analysis from Error Probability

Through the following diversity gain analysis, we can see the impact of each MIMO detector on the performance of multiuser systems.

6.1.1. Diversity Gain of Combinatorial User Selection with ML and MMSE Detectors

Using the pairwise error probability (PEP), we can find the diversity order from multiple receive antennas as well as multiple user selection.

Theorem 6.1. *The average PEP of the ML detector with the M selected users under the MDist user selection criterion, denoted by* P_e^{ml}, *is upper-bounded as*

$$P_e^{\text{ml}} \le c_1 \left(\frac{\|\sigma_h^2 \mathbf{d}\|^2}{N_0}\right)^{-N\lfloor \frac{K}{M} \rfloor} + o\left(\left(\frac{\|\sigma_h^2 \mathbf{d}\|^2}{N_0}\right)^{-N\lfloor \frac{K}{M} \rfloor + 1}\right), \tag{31}$$

where $c_1 > 0$ *is constant, and* $\mathbf{d} = \mathbf{s}_{(1)} - \mathbf{s}_{(2)}$ *(here,* $\mathbf{s}_{(i)} \in \mathcal{S}^{MP}$ *and* $\mathbf{s}_{(1)} \neq \mathbf{s}_{(2)}$*).*

Proof. See Section 8.1. □

This theorem shows that a full receive diversity gain of N together with a partial multiuser diversity gain of at least $\lfloor \frac{K}{M} \rfloor$ can be achieved by the ML detector under the MDist user selection criterion. This result is derived under the fact that there are at least $\lfloor \frac{K}{M} \rfloor$ statistically independent alternative combinations of the composite channel matrix $\mathbf{H}_{\mathcal{K}}$ for M users. Hence, this result is a lower bound on the diversity gain. In fact, there are more combinations for $\mathbf{H}_{\mathcal{K}}$, which are not independent, that can increase the multiuser diversity gain. By simulations, we will further demonstrate the impact of the combinations of M selected users that are not independent.

Theorem 6.2. *The average PEP of the MMSE detector with the selected M users under the ME user selection criterion, denoted by* P_e^{mmse}*, is upper-bounded as*

$$P_e^{\text{mmse}} \leq c_2 \left(\frac{\sigma_h^2 \|\mathbf{d}\|^2}{N_0} \right)^{-(N-P+1)\lfloor \frac{K}{M} \rfloor} + o\left(\left(\frac{\sigma_h^2 \|\mathbf{d}\|^2}{N_0} \right)^{-(N-P+1)\lfloor \frac{K}{M} \rfloor + 1} \right), \tag{32}$$

where $c_2 > 0$ *is constant.*

Proof. See Section 8.2. □

This theorem shows that for the MMSE detector, the ME user selection criterion may not be able to exploit a full receive diversity.

6.1.2. Diversity Gain of Combinatorial User Selection with LR-based Detector

Theorem 6.3. *The average PEP of the LR-based SIC detector with the selected M users under the MD user selection criterion, denoted by* P_e^{lr}*, is upper-bounded as*

$$P_e^{\text{lr}} \leq c_3 \left(\frac{\|\sigma_h^2 \mathbf{d}\|^2}{N_0} \right)^{-N\lfloor \frac{K}{M} \rfloor} + o\left(\left(\frac{\|\sigma_h^2 \mathbf{d}\|^2}{N_0} \right)^{-N\lfloor \frac{K}{M} \rfloor + 1} \right), \tag{33}$$

where $c_3 > 0$ *is constant.*

Proof. See Section 8.3. □

This theorem shows that a full receive diversity gain of N together with the same partial multiuser diversity gain, $\lfloor \frac{K}{M} \rfloor$, as with the ML detector, can be achieved by the LR-based detector under the MD user selection criterion. From these results, we can see that the LR-based detector is as good as the ML detector with respect to the diversity gains.

6.2. Numerical results

In this subsection, we present simulation results with the MIMO channels of $\sigma_h^2 = 1$. The SNR is defined by the energy per bit to the noise power spectral density ratio E_b/N_0. We used 16 quadrature amplitude modulation (16-QAM) for signaling with Gray mapping. CLLL is carried out for the lattice basis reduction.

6.2.1. Single user selection

In order to illustrate the impact of the diversity gain to multiuser MIMO systems, we first present the bit error rate (BER) performance of various multiuser MIMO systems in Fig. 1, where only a single user is selected at one time (i.e., $M = 1$). Five multiuser MIMO systems are considered with $P = N = 4$ and $K = 10$, namely:

1. MMSE detection under ME criterion: MMSE (ME).
2. ML detection under MDist criterion: ML (MDist).
3. LR-based MMSE-SIC detection under maximize mutual information (MMI) criterion: LR-based MMSE-SIC (MMI).
4. LR-based MMSE-SIC detecton under optimal decision region (ODR) criterion: LR-based MMSE-SIC (ODR)
5. LR-based MMSE-SIC detection under MD criterion: LR-based MMSE-SIC (MD)

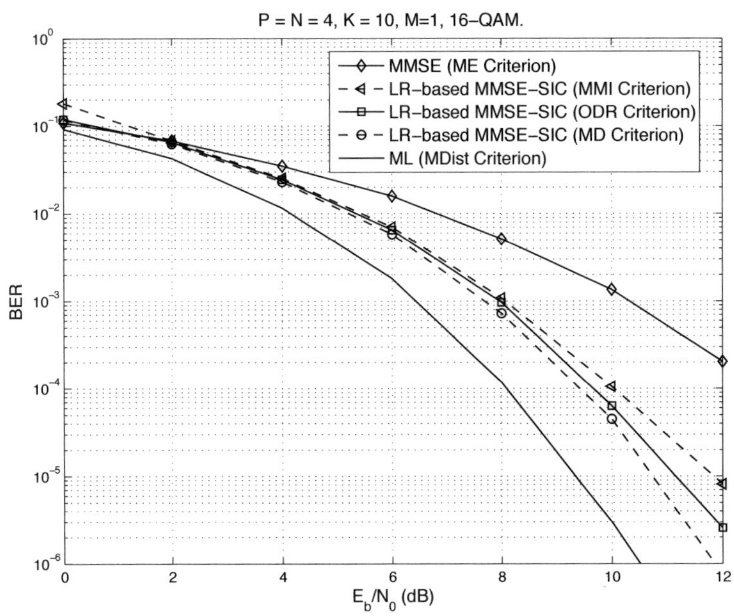

Figure 1. BER performance of various multiuser MIMO systems with 16-QAM, $P = N = 4$, $M = 1$, and $K = 10$.

From Fig. 1, we can observe that the optimal performance is guaranteed by the ML detection under MDist criterion. On the other hand, the conventional MMSE detector with the ME criterion provides poor performance as they cannot fully exploit spatial diversity.

Alternatively, the LR-based SIC detector with the MD criterion can exploit a full diversity as the ML detector with the MDist criterion. It it noteworthy that a full diversity gain cannot be achieved by the LR-based MMSE-SIC detection with the MMI criterion, although the performance can be improved by using the ODR criterion, there is still a BER gap compared to the one with the MD criterion. Overall, it is shown that the best user selection criterion for the LR-based MMSE-SIC detection is the MD criterion.

6.2.2. Multiple users selection

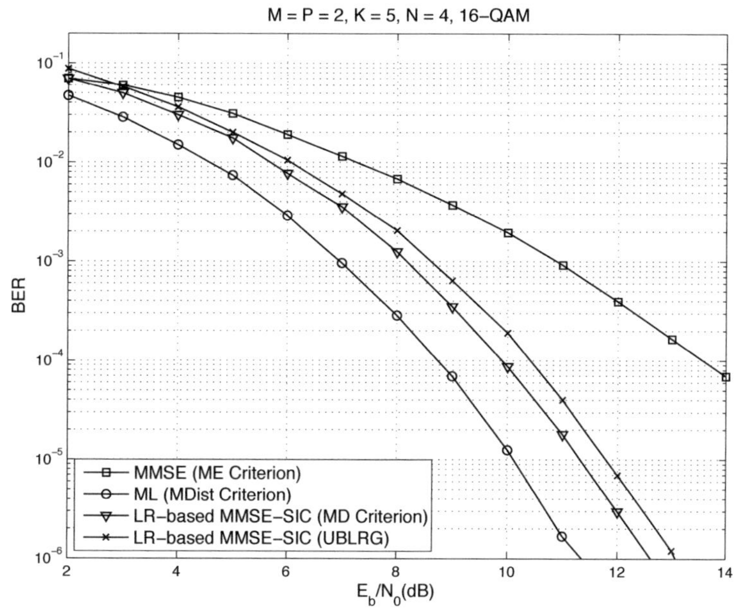

Figure 2. BER performance of various multiuser MIMO systems with 16-QAM, $M = P = 2$, $N = 4$, and $K = 5$.

To see the performance of different multiple users selection criteria, the BER results are shown in Fig. 2 for the case of $M = P = 2$. We assume that $K = 5$ and $N = 4$. It is shown that, when BER drops from 10^{-5} to 10^{-6}, SNR increases by approximately 1.2 dB. Thus, an estimate of the diversity gain from the simulation becoms $G \simeq 8.3$, which is greater than the lower bound, $G_{low} = N\lfloor K/M \rfloor = 8$, derived from the theoretical analysis in Section 6.1. Moreover, it is shown that the user selection approach with the LR-based detectors has the same diversity gain as in the ML detector, whereas the approach with the MMSE detector has a lower diversity gain. In general, we can show that system of LR-based MMSE-SIC detector with UBLRG can provide a reasonably good performance. Note that, compared with the LR-based MMSE-SIC detector with MD criterion and combinatorial selection, the proposed UBLRG system provides a similar performnace; however, as shown in Table 2, by decreasing the number of column swapping, complexity can also be reduced remarkably with more efficient implementations for the proposed UBLRG approach.

7. Conclusion

In this chapter, we studied the user selection based on the error probability of an actually employed MIMO detector in multiuser MIMO systems. As the complexity becomes prohibitively high if the user selection is based on exhaustive search (i.e., the combinatorial user selection), we considered a greedy user selection approach to keep the complexity low. We showed that low-complexity suboptimal detectors (i.e., the LR-based MMSE-SIC detector) with the MD criterion for the user selection can fully exploit both multiuser and receive diversity and provide good performance even though their complexity is low, which has been confirmed by both theoretical analysis and simulation results. Moreover, according to the simulation results, it was also shown that the LR-based detection with our proposed greedy user selection approach can achieve a similar diversity gain and have a comparable performance with that based on a combinatorial approach.

8. Appendix

8.1. Proof of Theorem 6.1

Proof. With the selected M users by the combinatorial user selection approach under the MDist criterion, suppose that we jointly detect M users' signals with the $N \times MP$ channel matrix $\mathbf{H}_{\mathcal{K}}$ using the ML detector. The PEP in detecting M users' signals has the following upper bound:

$$\Pr\left(\mathbf{s}_{(1)} \to \mathbf{s}_{(2)}\right) \leq \operatorname{erfc}\left(\sqrt{\frac{\|\mathbf{H}_{\mathcal{K}}\bar{\mathbf{d}}\|^2}{2N_0}}\right), \tag{34}$$

where

$$\bar{\mathbf{d}} = \arg \min_{\mathbf{d}\in\mathcal{D},\mathbf{d}\neq 0} \|\mathbf{H}_{\mathcal{K}}\mathbf{d}\|^2,$$

$$\mathcal{D} = \left\{\mathbf{d} = \mathbf{s} - \mathbf{s}' \mid \mathbf{s} \neq \mathbf{s}' \in \mathcal{S}^{MP}\right\} \subset \mathbb{Z}^{MP} + j\mathbb{Z}^{MP}, \tag{35}$$

and $\operatorname{erfc}(x)$ is the complementary error function of x, i.e., $\operatorname{erfc}(x) = \frac{2}{\sqrt{\pi}} \int_x^{+\infty} e^{-z^2} dz$.

Let $\mathcal{V}(\mathbf{H}_{\mathcal{K}})$ denote the length of the shortest non-zero vector of the lattice generated by $\mathbf{H}_{\mathcal{K}}$. Then, we have

$$\Pr\left(\mathbf{s}_{(1)} \to \mathbf{s}_{(2)}\right) \leq \operatorname{erfc}\left(\sqrt{\frac{\mathcal{V}^2(\mathbf{H}_{\mathcal{K}})}{2N_0}}\right), \tag{36}$$

where

$$\mathcal{V}(\mathbf{H}_{\mathcal{K}}) = \|\mathbf{H}_{\mathcal{K}}\bar{\mathbf{d}}\|. \tag{37}$$

http://dx.doi.org/10.5772/57130

For the case that the MDist criterion is employed, we have

$$\Pr\left(\mathbf{s}_{(1)} \to \mathbf{s}_{(2)}\right) \leq \operatorname{erfc}\left(\sqrt{\frac{\max_{\mathcal{K}} \mathcal{V}^2(\mathbf{H}_{\mathcal{K}})}{2N_0}}\right), \tag{38}$$

Note that

$$\max_{\mathcal{K}} \mathcal{V}^2(\mathbf{H}_{\mathcal{K}}) = \max_{\mathcal{K}} \min_{\mathbf{d}\in\mathcal{D},\mathbf{d}\neq 0} \mathbf{d}^{\mathrm{H}}\mathbf{H}_{\mathcal{K}}^{\mathrm{H}}\mathbf{H}_{\mathcal{K}}\mathbf{d}, \tag{39}$$

Let $\mathbf{w}_{\mathcal{K}} = \mathbf{H}_{\mathcal{K}}\mathbf{d}$. Note that $\mathbf{w}_{\mathcal{K}}$ is a zero-mean CSCG random vector and

$$E\left[\mathbf{w}_{\mathcal{K}}\mathbf{w}_{\mathcal{K}}^{\mathrm{H}}\right] = \sigma_h^2\|\mathbf{d}\|^2\mathbf{I}. \tag{40}$$

We can show that $X_{\mathcal{K}} = ||\mathbf{w}_{\mathcal{K}}||^2$ is a chi-square random variable with $2N$ degrees of freedom and its pdf is

$$f_X(x_{\mathcal{K}}) = \frac{1}{(\sigma_h^2\|\mathbf{d}\|^2)^N(N-1)!} x_{\mathcal{K}}^{N-1} e^{-x_{\mathcal{K}}/(\sigma_h^2\|\mathbf{d}\|^2)}. \tag{41}$$

The cumulative distribution function (cdf) is

$$F_X(x_{\mathcal{K}}) = 1 - e^{x_{\mathcal{K}}/(\sigma_h^2\|\mathbf{d}\|^2)} \sum_{q=0}^{N-1} \frac{(x_{\mathcal{K}}/(\sigma_h^2\|\mathbf{d}\|^2))^q}{q!}. \tag{42}$$

To obtain an upper bound on the error probability, we note that the number of alternative combinations of the channel matrices, which are statistically independent with each other, for selecting $\mathbf{H}_{\mathcal{K}}$ with the MDist selection is at least $\lfloor \frac{K}{M} \rfloor$. Let $\mathbf{H}_{\mathcal{K}_1}, \mathbf{H}_{\mathcal{K}_2}, \ldots, \mathbf{H}_{\mathcal{K}_{\lfloor \frac{K}{M} \rfloor}}$ represent such $\lfloor \frac{K}{M} \rfloor$ independent alternative combinations of the channel vectors. Then, there are at least $\lfloor \frac{K}{M} \rfloor$ of $\mathbf{w}_{\mathcal{K}}$, i.e., $\mathbf{w}_{\mathcal{K}_1}, \mathbf{w}_{\mathcal{K}_2}, \ldots, \mathbf{w}_{\mathcal{K}_{\lfloor \frac{K}{M} \rfloor}}$, which are independent. Let $V = \max\left\{X_1, X_2, \ldots, X_{\lfloor \frac{K}{M} \rfloor}\right\}$, where $X_m = ||\mathbf{w}_{\mathcal{K}_m}||^2$. Using order statistics, the pdf of V is given by

$$f_V(v) = K F_X^{\lfloor \frac{K}{M} \rfloor - 1}(v) f_X(v) = c_1' v^{N\lfloor \frac{K}{M} \rfloor - 1} + o(v^{N\lfloor \frac{K}{M} \rfloor - 1 + \epsilon}), \tag{43}$$

where $c_1' > 0$ is a constant, and $\epsilon > 0$. Thus, we have

$$\begin{aligned} P_e^{\mathrm{ml}} &\leq \sum_{\mathbf{d}\in\mathcal{D},\mathbf{d}\neq 0} E_V\left[\operatorname{erfc}\left(\sqrt{\frac{\max_{\mathcal{K}} \mathbf{d}^{\mathrm{H}}\mathbf{H}_{\mathcal{K}}^{\mathrm{H}}\mathbf{H}_{\mathcal{K}}\mathbf{d}}{2N_0}}\right)\right] \\ &= c_1\left(\frac{\|\sigma_h^2\mathbf{d}\|^2}{N_0}\right)^{-N\lfloor \frac{K}{M} \rfloor} + o\left(\left(\frac{\|\sigma_h^2\mathbf{d}\|^2}{N_0}\right)^{-N\lfloor \frac{K}{M} \rfloor + 1}\right), \end{aligned} \tag{44}$$

where $c_1 > 0$ is a constant. This completes the proof. □

8.2. Proof of Theorem 6.2

Proof. It can be shown that under the ME criterion, for a given $\mathbf{H}_{\mathcal{K}}$, an upper bound on the error probability in detecting M users' signals is expressed as

$$P_e^{\text{mmse}} \leq \text{erfc}\left(\sqrt{\frac{\max_{\mathcal{K}} \lambda_{\min}(\mathbf{H}_{\mathcal{K}}^{\text{H}}\mathbf{H}_{\mathcal{K}})||\mathbf{d}||^2}{2N_0}}\right)$$
$$= \text{erfc}\left(\sqrt{\frac{\sigma_h^2||\mathbf{d}||^2 \max_{\mathcal{K}} \tilde{X}_{\mathcal{K}}}{2N_0}}\right)$$
$$= \text{erfc}\left(\sqrt{\frac{\sigma_h^2||\mathbf{d}||^2 V}{2N_0}}\right), \tag{45}$$

where $\tilde{X}_{\mathcal{K}} = \lambda_{\min}(\mathbf{H}_{\mathcal{K}}^{\text{H}}\mathbf{H}_{\mathcal{K}})/\sigma_h^2$ and $V = \max_{\mathcal{K}} \tilde{X}_{\mathcal{K}}$.

Using the pdf of V (with the same derivation for the ML case in the last subsection), it can be deduced that

$$P_e^{\text{mmse}} = E_{\mathbf{H}_{\mathcal{K}}}[\Pr\left(\mathbf{s}_{(1)} \rightarrow \mathbf{s}_{(2)}\right)]$$
$$\leq E_V\left[\text{erfc}\left(\sqrt{\frac{\sigma_h^2||\mathbf{d}||^2 V}{2N_0}}\right)\right]. \tag{46}$$

For independent alternative combinations of the channel matrices $\mathbf{H}_{\mathcal{K}_1}, \mathbf{H}_{\mathcal{K}_2}, \ldots, \mathbf{H}_{\mathcal{K}_{\lfloor \frac{K}{M} \rfloor}}$, similar to the proof of Theorem 5.1, we can obtain that

$$P_e^{\text{mmse}} \leq E_V\left[\text{erfc}\left(\sqrt{\frac{\sigma_h^2||\mathbf{d}||^2 V}{2N_0}}\right)\right]$$
$$\leq \int_0^{+\infty} \text{erfc}\left(\sqrt{\frac{\sigma_h^2||\mathbf{d}||^2 v}{2N_0}}\right) f_V(v)dv$$
$$= c_2\left(\frac{\sigma_h^2||\mathbf{d}||^2}{N_0}\right)^{-(N-P+1)\lfloor \frac{K}{M} \rfloor} + o\left(\left(\frac{\sigma_h^2||\mathbf{d}||^2}{N_0}\right)^{-(N-P+1)\lfloor \frac{K}{M} \rfloor + 1}\right), \tag{47}$$

where $c_2 > 0$ is constant. This completes the proof. □

8.3. Proof of Theorem 6.3

Proof. In the LR algorithm, we transform the given channel matrix, e.g., $\mathbf{H}$, into a new basis, e.g., denoted by $\mathbf{G}$. Here, we have $\mathcal{L}(\mathbf{G}) = \mathcal{L}(\mathbf{H}) \Longleftrightarrow \mathbf{G} = \mathbf{HT}$, where $\mathbf{T}$ is an integer unimodular matrix and $\mathcal{L}(\mathbf{A})$ denotes the lattice generated by $\mathbf{A}$. Then, $\mathbf{G}$ is called LLL-reduced with parameter δ if $\mathbf{G}$ is QR factorized as $\mathbf{G} = \mathbf{QR}$ where $\mathbf{Q}$ is unitary, $\mathbf{R}$ is upper triangular, and the elements of $\mathbf{R}$ satisfies (29) and (30) with $m = M$. We rewrite (30) as

$$\delta \mid r_{\rho,\rho} \mid^2 \leq \mid r_{\rho,\rho+1} \mid^2 + \mid r_{\rho+1,\rho+1} \mid^2, \ \ \rho = 1, 2, \ldots, MP - 1. \tag{48}$$

Then, we can obtain the following inequalities:

$$\mid r_{\rho+1,\rho+1} \mid^2 \geq \beta^{-1} \mid r_{\rho,\rho} \mid^2, \tag{49}$$

where $\beta = (\delta - \frac{1}{4})^{-1} > \frac{4}{3}$, and

$$\min_{\rho} \mid r_{\rho,\rho} \mid^2 \geq \beta^{-MP+1} \mid r_{1,1} \mid^2 . \tag{50}$$

Since $\mathbf{G} = \mathbf{QR}$, we have $\mid r_{1,1} \mid^2 = \|\mathbf{g}_1\|^2$ and

$$\|\mathbf{g}_1\|^2 \geq \min_{\mathbf{d} \in \mathcal{D}, \mathbf{d} \neq 0} \|\mathbf{Hd}\|^2 = \mathcal{V}^2(\mathbf{H}). \tag{51}$$

Thus, we have

$$\min_{\rho} \mid r_{\rho,\rho} \mid^2 \geq \beta^{-MP+1} \mathcal{V}^2(\mathbf{H}). \tag{52}$$

In the proposed user selection for selecting M users with the LR-based SIC detectors, (52) becomes

$$\min_{\rho} \mid r_{\rho,\rho} \mid^2 \geq \beta^{-MP+1} \mathcal{V}^2(\mathbf{H}_{\mathcal{K}}), \tag{53}$$

where $\mathcal{K}$ is the index set of the selected users.

Note that the LR-based SIC detection is considered. Let n_ρ denote the ρth element of $\tilde{\mathbf{n}}$. Then, the LR-based SIC detection does not have error across all the layers if we have $\frac{|n_\rho|}{|r_{\rho,\rho}|} < \frac{1}{2}$ or $|n_\rho|^2 < \frac{|r_{\rho,\rho}|^2}{4}$ for all ρ. Thus, the error probability of the LR-based SIC detector can be estimated by

$$\Pr(\text{error}) \simeq \exp\left(- \min_{q} \frac{|r_{\rho,\rho}|^2}{4N_0} \right). \tag{54}$$

Note that the approximation in above becomes accurate as $N_0 \to 0$ (or high SNR).

Substituting (53) into (54), we have

$$\begin{aligned}\Pr(\text{error}) &\leq \exp\left(-\beta^{-MP+1}\mathcal{S}^2(\mathbf{H}_{\mathcal{K}})\right) \\ &\leq \sum_{\mathbf{d}\in\mathcal{D},\mathbf{d}\neq 0} \exp\left(-\beta^{-MP+1}\frac{\max_{\mathcal{K}} \mathbf{d}^H\mathbf{H}_{\mathcal{K}}^H\mathbf{H}_{\mathcal{K}}\mathbf{d}}{2N_0}\right).\end{aligned} \tag{55}$$

Then, with the same approach used in the proof of Theorem 5.1, we can show that the upper bound on the average PEP is

$$P_e^{\mathrm{lr}} \leq c_3\left(\frac{\|\sigma_h^2\mathbf{d}\|^2}{N_0}\right)^{-N\lfloor\frac{K}{M}\rfloor} + o\left(\left(\frac{\|\sigma_h^2\mathbf{d}\|^2}{N_0}\right)^{-N\lfloor\frac{K}{M}\rfloor+1}\right), \tag{56}$$

where $c_3 > 0$ is constant. This completes the proof. □

Acknowledgements

This work was supported by the National Natural Science Foundation of China (Grant No. 61231011, No. 61231013, and No. 61250001).

Author details

Qiaoyu Li[1,*], Ying Li[2], Lin Bai[1] and Jinho Choi[3]

* Address all correspondence to: q.li@ee.buaa.edu.cn

1 School of Electronic and Information Engineering, Beihang University, Beijing, China
2 Institute of China Electronic Equipment System Engineering Corporation, Beijing, China
3 School of Information and Communications, Gwangju Institute of Science and Technology (GIST), Gwangju, Korea

References

[1] A J Paulraj and C B Papadias. Space-time processing for wireless communications. *IEEE Dignal Processing Magazine*, 14(6):49–83, Nov. 1997.

[2] L C Godara. Applications of antenna arrays to mobile communications-part i: performance improvement, feasibility, and system considerations. *Proceedings of the IEEE*, 85(7):1031–1060, Jul. 1997.

[3] L C Godara. Applications of antenna arrays to mobile communications-part ii: beam-forming and direction-of-arrival considerations. *Proceedings of the IEEE*, 85(8):1195–1245, Aug. 1997.

[4] A K Lenstra, H W Lenstra, and L Lovász. Factoring polynomials with rational coefficients. *Mathematische Annalen*, 261(4):515–534, Dec. 1982.

[5] C Windpassinger, L H J Lampe, and R F H Fischer. From lattice-reduction-aided detection towards maximum-likelihood detection in mimo systems. In *Proceedings of IEEE Information Theroy Workshop*, pages 144–148, Mar. 2003.

[6] H Yao and G W Wornell. Lattice-reduction-aided detectors for mimo communication systems. In *Proceedings of IEEE Global Telecommunications Conference*, pages 424–428, Nov. 2002.

[7] D Wübben, R Böhnke, V Kühn, and K D Kammeyer. Near-maximum-likelihood detection of mimo systems using mmse-based lattice reduction. In *Proceedings of IEEE International Conference on Communications*, pages 798–802, Jun. 2004.

[8] Y H Gan, C Ling, and W H Mow. Complex lattice reduction algorithm for low-complexity full-diversity mimo detection. *IEEE Transactions on Signal Processing*, 57(7):2701–2710, Jul. 2004.

[9] R Knopp and P Humblet. Information capacity and power control in single-cell multiuser communications. In *Proceedings of IEEE International Conference on Communications*, pages 331–335, Jun. 1995.

[10] M Bengtsson. From single-link mimo to multiuser mimo. In *Proceedings of IEEE International Conference on Acoustics, Speech, and Signal Processing*, volume 4, pages 697–700, May 2004.

[11] G Dimic and N Sidiropopulos. On downlink beamforming with greedy user selection: performance analysis and a simple new algorithm. *IEEE Transactions on Signal Processing*, 53(10):3857–3868, Oct. 2005.

[12] R W Heath, Jr S Sandhu, and A J Paulraj. Antenna selection for spatial multiplexing systems with linear receivers. *IEEE Communications Letters*, 5(4):142–144, Apr. 2001.

[13] I Berenguer and X Wang. Mimo antenna selection with lattice-reduction-aided linear receivers. *IEEE Transactions on Vehicular Technology*, 53(5):1289–1302, Sep. 2004.

[14] T Yoo, N Jindal, and A Goldsmith. Multiantenna broadcast channels with limited feedback and user selection. *IEEE Journal on Selected Areas in Communications*, 25(7):1478–1491, Sep. 2007.

[15] A Gorokhov, D A Gore, and A J Paulraj. Reveive antenna selection for mimo flat-fading channels: theory and algorithms. *IEEE Transactions on Information Theory*, 49(10):2796–2807, Oct. 2003.

[16] A Gorokhov, D A Gore, and A J Paulraj. Reveive antenna selection for mimo spatial multiplexing: theory and algorithms. *IEEE Transactions on Signal Processing*, 51(11):2796–2807, Nov. 2003.

[17] M Gharavi-Alkhansari and A B Gershman. Fast antenna subset selection in mimo systems. *IEEE Transactions on Signal Processing*, 52(2):339–347, Feb. 2004.

[18] A Bayesteh and A K Khandani. On the user selection for mimo broadcast channels. *IEEE Transactions on Information Theory*, 54(3):1086–1107, Mar. 2008.

[19] Z Shen, R Chen, J G Andrews, and R W Heath. Low-complexity user selection algorithms for multiuser mimo systems with block diagonalization. *IEEE Transactions on Signal Processing*, 54(9):3658–3663, Sep. 2006.

[20] T Ji, C Zhou, S Zhou, and Y Yao. Low-complexity user selection strategies for multiuser mimo downlink scenario. In *Proceedings of IEEE Wireless Communications and Networking Conference*, pages 1532–1537, Mar. 2007.

[21] S Lee and J S Thompson. Qos-guaranteed sequential user selection in multiuser mimo downlink channels. In *Proceedings of IEEE Vehicular Technology Conference*, pages 1926–1930, Apr. 2007.